D0938298

The Gene-Splicing Wars

AAAS Issues in Science and Technology Series

Science as Intellectual Property:
Who Controls Research?
Dorothy Nelkin

Scientists and Journalists:
Reporting Science as News
Sharon M. Friedman, Sharon Dunwoody,
and Carol Rodgers, *Editors*

Low Tech Education in a High Tech World:
Corporations and Classrooms
in the New Information Society
Elizabeth L. Useem

Science and Creation:
Geological, Theological, and
Educational Perspectives
Robert W. Hanson, *Editor*

The Gene-Splicing Wars:
Reflections on the Recombinant DNA Controversy
Raymond A. Zilinskas and Burke K. Zimmerman, *Editors*

The Gene-Splicing Wars

Reflections on the Recombinant DNA Controversy

Edited by
RAYMOND A. ZILINSKAS
and
BURKE K. ZIMMERMAN

Issues in Science and Technology Series
American Association for the Advancement of Science

MACMILLAN PUBLISHING COMPANY
A Division of Macmillan, Inc.
NEW YORK
COLLIER MACMILLAN PUBLISHERS
LONDON

This volume grew out of a preliminary meeting held at the University of Southern California in April 1981 and a subsequent symposium at the 1982 Annual Meeting of the American Association for the Advancement of Science. For their help in bringing about that meeting, the authors and editors wish to thank Dr. Robert Friedheim and the Institute for Marine and Coastal Studies at USC. Financial support for the first meeting and to ensure that all contributors could attend the AAAS symposium was generously provided by the Ahmanson Foundation, Cetus Corporation, the Los Angeles Pasteur Library, and the USC School of Liberal Arts. The editors and authors wish to thank Dr. Ron Cape, Dr. Preston Dent, Dr. Norman Kharasch, and Dr. Norman Topping, all of whom were instrumental in generating this support.

The efficient and energetic secretarial and organizational help of Ms. Ellen Kinsinger of USC during the early phases of the project and of Ms. Eloise Sanchez of Cetus Corporation during the later phases is gratefully acknowledged. Thanks are also due to Mr. Eric Ladner and the Cetus Graphics Department for producing appendix figures and to Plenum Publishing Corporation for their permission to reproduce them in this volume.

Finally, thanks are due to Dr. Kathryn Wolff, Ms. Joellen Fritsche and Ms. Sue O'Connell of the AAAS Meetings and Publications Department for their editorial assistance in the final phases of the project.

Macmillan Publishing Company
866 Third Avenue, New York, NY 10022

Collier Macmillan Canada, Inc.

Printed in the United States of America

printing number 2 3 4 5 6 7 8 9 10
year 7 8 9 0 1 2 3 4 5

Library of Congress Cataloging-in-Publication Data
Main entry under title:

The Gene-splicing wars.

(AAAS series on issues in science and technology)
Includes index.
1. Recombinant DNA—Social aspects. 2. Recombinant DNA—Moral and ethical aspects. 3. Genetic engineering—Social aspects. 4. Genetic engineering—Moral and ethical aspects. I. Zilinskas, Raymond A. II. Zimmerman, Burke K. III. Series
QH442.G438 1985 174'.9574 84-29446
ISBN 0-02-948560-6

About The Authors

Donald S. Fredrickson, M.D.
President and Chief Executive Officer, Howard Hughes Medical Institute, Washington, D.C.

Keith Gibson, Ph.D.
Deputy Chief Scientific Officer, Medical Research Council, London, U.K.

Clifford Grobstein, Ph.D.
Professor of Biological Sciences and Public Policy, Program on Science, Technology, and Public Policy, University of California, San Diego, La Jolla, California

Harlyn O. Halvorson, Ph.D.
Director, Rosenstiel Basic Medical Sciences Research Center, Brandeis University, Waltham, Massachusetts

Zsolt Harsanyi, Ph.D.
President, Porton International Inc., Washington, D.C.

Sheldon Krimsky, Ph.D.
Associate Professor, Department of Urban and Environmental Policy, Tufts University, Medford, Massachusetts

Donald N. Michael, Ph.D.
Consultant in futures issues and technology policy, San Francisco, California

Claire Nader, Ph.D.
Science Policy Consultant, Washington D.C.

Nanette Newell, Ph.D.
Consultant, Allelix, Inc., Mississauga, Ontario, Canada

Harold Schmeck, Jr.
Science Correspondent, *New York Times*, New York, New York

Ray Thornton
President, University of Arkansas, Fayetteville, Arkansas

Raymond A. Zilinskas, Ph.D.
Science Policy Consultant, Santa Monica, California

Burke K. Zimmerman, Ph.D.
Adjunct Professor, Department of Medicine, Division of Medical Ethics, University of California, San Francisco, and Visiting Professor, Graduate School of Public Policy, University of California, Berkeley

Norton D. Zinder, Ph.D.
John D. Rockefeller, Jr., Professor of Molecular Genetics, Rockefeller University, New York, New York

Contents

Preface

One day Chicken Little was playing in her yard when an acorn from a tall oak fell and hit her on the head.

"The sky is falling," cried Chicken Little. "We must rush to tell the king."

She then ran to tell Henny Penny, Ducky Lucky, Goosey Lucy, and Turkey Lurkey, each in turn, the alarming news that the sky was falling. They all got quite excited, not questioning for a second the veracity of their trusted colleague's proclamation. Off they all went, rushing to tell the king. On the way, they met the sly Foxy Loxy who preyed on their fears and, ultimately, on their succulent bodies.

Except possibly for the last sentence, the familiar moral nursery tale could be considered an apt allegory of what is politely described as the recombinant DNA controversy, but for which the editors believe the "gene-splicing wars" would be a much more accurate description.

Looking back on the events of the late 1970s from a point several years removed, we see much more than a controversial issue in science, or science policy, but something close to a laboratory experiment in human social psychology—an experiment in the responses to fear. When viewed in that way, the wisdom of the parable of Chicken Little is inescapable and so has provided the editors with an appropriate theme for this volume. This analogy, of course, raises a sensitive question: Who was Chicken Little?

To that question, there are many answers, all correct. The cries of imminent catastrophe came from many quarters. First, of course, there were the fears of those who were anxious about the ominous overtones of a complex new technology that they did not really understand. They rushed to tell the king to stop the sky from falling. Then there were those who feared a legislative or regulatory calamity might befall the scientific research enterprise. Lab coats flying, they rushed en masse to tell the king of the terrible fate about to come down on their heads. Only some were not quite sure that the king was the right one to tell. If the U.S. capitol was the king's palace, many would have preferred to take their chances with Foxy Loxy. And then there were others, defending the rights of the people, who saw disaster threatening from all directions—the labs, the corporations, the Congress. They too rushed to tell the king, and to lobby him for salvation. Perhaps there were more.

But the sky never fell. Chicken Little was apparently wrong. And the great flurry of everyone stumbling over each other in their rush to tell the king eventually dissipated. Foxy Loxy fell on hard times, while the Chicken Littles of the world went back to their homes and offices and laboratories and corporate headquarters and seats on legislative bodies.

Foxy Loxy, whose livelihood depends on mass hysteria, is occasionally seen filing suit against the federal government or trying to get people to believe that the genetic heritage of a cow is an inviolable and sacred right. But now, in the mid-1980s, it is hard to find a Chicken Little or a Henny Penny or a Ducky Lucky.

This volume is the story of the key events of those turbulent days viewed from many different perspectives by a unique group of individuals, each of whom was a protagonist in the drama of the gene-splicing wars in some major way. They examine why some thought the sky was falling and how they tried to stop it. They look into the past to see if there are lessons to be learned and, more important, if society is capable of learning them. And what of the future? Could it all happen again? Some still have their heads cocked skyward, watching and waiting. They ask, Was Chicken Little right after all?

BURKE K. ZIMMERMAN

Introduction

During the early 1970s a new term was added to our vocabulary: "recombinant DNA." At first it simply described a new technique that facilitated the study of gene function. But as it became the focal point in a controversy that pitted research scientists, the U.S. Congress, public interest groups, corporations, and citizens against each other in a multidimensional free-for-all, it acquired a spectrum of identities. It symbolized the brilliant wonders of science and technology; it also symbolized the ominous and fiendish creations of the stereotypical mad scientist. Newspaper cartoons, movies, and television dramas capitalized on all of the dazzling and creepy overtones of meaning that "recombinant DNA" conjured up, further distorting the public image of this simple technique of molecular genetics while at the same time attributing to it a moral identity and value as well as powers and properties that far exceeded reality.

The first concerns, originally raised by the scientist-inventors of the method, were over safety and the potential hazards of combining the genes of one organism with those of another. There were insufficient data to answer questions of risk on scientific grounds, but the perceptions of risk varied over the widest possible range. Since 1973, when the matter of risk was first publicly raised, these perceptions have changed greatly, somewhat out of proportion to the new data on risk actually gained. The experience of more than a decade of research without discernible injury has given comfort to both scientists and laymen. The scientists who first raised the question are no longer worried, despite the field having proliferated a thousandfold and the widespread deliberate modification of the genes of a vast array of organisms, not just the bacterium *Escherichia coli* (of the human gut), the workhorse of the early years. Why?

Of course, this technique, along with some other significant advances in the analytical methods in molecular biology, has revolutionized research. We now have the ability to clarify the fundamental relationships between the coded molecular information in DNA and the ultimate expression of every aspect of life. And we are now, or soon will be, at a point where we are able to intervene by design in the genetic makeup of all species, create new ones, and modify the once sacred "human" in perhaps undreamed of ways.

The fires that raged in the mid- to late 1970s over recombinant DNA have left but a few glowing embers. What fueled the fire—fear? ignorance? lack of trust in science? What set it off, and why—with so many questions still unanswered—did it seem to burn itself out?

In this volume, we try to present the reader with new insights into the explosive recombinant DNA debate. All of the authors were involved in the action—most of them directly. But their viewpoints are very different, often, so are their conclusions. As a lesson in learning about ourselves, we believe the gene-splicing wars have taught us something about human behavior that will help us to deal with our future. Perhaps we are now better able to predict when new issues will be greeted with a similar response, or when equally significant advances will scarcely be noticed.

While "recombinant DNA" may now quietly reside in the lexicon of molecular biologists as only a method, its legacy is revolutionary, for it allows us to observe the elegance, beauty, and extraordinary ingenuity of some of nature's hitherto most carefully guarded secrets. Inseparable from the acquisition of this profound knowledge is the awesome responsibility to respect it accordingly. Can and will we use this knowledge with wisdom? How should we make the difficult decisions that lie ahead? Do we run the risk of setting off more flames of controversy, or of refueling the hot residue of the last one? These issues are also addressed in this volume.

We asked our contributors not simply to reflect upon the past in order to gain perspective on the phenomenon of controversy itself and on human nature as mankind tries to deal with a world that is changing, perhaps too fast. We also asked them to look to the future. How shall man deal with new knowledge of profound implications? How can risk be realistically estimated for phenomena that are only beginning to be understood? What determines the perceptions of risk and safety, and what causes these perceptions to change, even when data are not available to resolve the question on scientific grounds? What are the implications of a well-financed effort to commercialize an area of technology so intrinsically involved with basic research that they cannot be separated? How shall we resolve the ethical questions concerning coming applications of new discoveries that may soon allow us to intervene in genetic disease, alter the genetic makeup of our children, and perhaps slow or arrest the process of aging? Can our educational, commercial, professional, and political institutions deal with all of these questions rationally and effectively? Our contributors give their carefully considered opinions on these and other questions—opinions which, in spite of their differences, show at times a remarkable convergence.

In the first chapter in this volume, Clifford Grobstein provides an overall definition of the gene-splicing controversy, putting it in perspective as a public policy issue. He traces the roots of the controversy from Asilomar in 1975 and questions whether the actions of the scientific community were an appropriate response to the recognized possibility that deliberate genetic mixing could produce undesirable consequences.

Donald S. Fredrickson saw the controversy from his unique position as director of the National Institutes of Health. There he became the guardian of the oracle known as the NIH guidelines, as the antagonists in the debate challenged them for being too stringent or too permissive. Now freed from the constraints that bind a public servant, he shares with us his experiences as the NIH guidelines went through their creation and development, and his reflections on these years.

Former Congressman Ray Thornton, now president of the University of Arkansas, gives a rare view of the concerns that he and his subcommittee had about the implications of rDNA research as an issue in science policy and whether mandatory federal safety rules were needed. He later became chairman of the Recombinant DNA Advisory Committee (RAC) at the NIH. He comments on and augments Donald Fredrickson's views from the perspective of his own unique role in the controversy.

During the years of peak intensity of the controvesy, 1977–1978, Burke Zimmerman was responsible for devising appropriate and rational federal legislation to extend the NIH safety standards beyond the federal research sector. Thus, he observed the most frantic and strident aspects of the controversy from ground zero, for it was Congress that became the focal point of some of the most contentious issues. He continued his involvement in federal policy on rDNA as a special assistant to Donald Fredrickson at the NIH for the next two-and-a-half years.

The debate was perhaps most intense in the United States, but the issues and concerns that faced this country were also evident around the world. A view from abroad is presented by Keith Gibson, Secretary of the Genetic Manipulation Advisory Group (GMAG), at that time the British counterpart to RAC. As part of his work, he provided liaison with the several international organizations whose activities impinged on UK research activity, particularly the European Economic Community, as well as the European and American counterparts to GMAG.

In the present world where science and technology are major political issues, professional societies must extend their concerns far beyond the traditional activities of promoting a particular branch of science and research. One of the important consequences of the controversy was the emergence of scientists as organized lobbyists for the first time since the late 1940s, when the scientific community lobbied for the civilian control of atomic energy. The past president of the American Society for Microbiology, Harlyn Halvorson, served much of his term during the height of the rDNA controversy, and here he addresses that new role of scientists.

It has become a truism that a people who mean to be their own governors must arm themselves with the power of knowledge. The public's knowledge of science comes almost exclusively via the press, which can choose to be responsible or sensational. Many believe the rDNA controversy was given its emotional and contentious beginnings by a small number of highly sensationalized and inaccurate articles in the press. This, and the general responsibility and competence of the press in reporting scientific events accurately, is discussed by Harold Schmeck, Jr., a science writer for the *New York Times* who has observed and written about recombinant DNA for more than a decade.

Norton Zinder, a molecular biologist at the Rockefeller University, was an active participant in the recombinant DNA debate from the scientific community, writing countless letters to the NIH, to congressmen, and to his fellow scientists, attempting to alter the course of policy on this crucial issue. He shares with us his perceptions of what the important determinants in the recombinant DNA war really were.

Surely the paramount problem in government or in the determination of policy at any level is who decides who will make the decisions. Donald Michael, a distinguished social psychologist, explores how policy should ultimately be determined, while Claire Nader, a science policy consultant, views the controversy from the perspective of the public interest organizations. She tells us how these organizations perceived and defended the rights of the public to determine the course of science and technology and especially the right to define for itself how it should be protected from involuntary risks, however uncertain they may be.

Some issues that surfaced during the early days of the rDNA debate were obscured by immediate concerns over laboratory safety, potential risks to local areas, and so on. These issues have now become more prominent. Raymond Zilinskas, at the time of writing an analyst with the Office of Technology Assessment, discusses the international implications posed by rDNA research by focusing on the use of rDNA techniques for warfare or terrorist purposes. He worries that the technology may present to national decision-makers new powerful methods in biological warfare.

Zsolt Harsanyi, project director of the OTA study *Impacts of Applied*

Genetics and now at Porton International, examines the emerging industrialization of molecular biology and looks into its future while taking into account its widest implications on human society.

In the final chapter, a more general view of emerging issues that may pose problems to society is taken by Sheldon Krimsky. As a member of the Cambridge Experimentation Review Board and, subsequently, of the RAC, he gives his view of earlier events and what the future may hold. As we move into the "age of genetic alchemy," Krimsky finds it imperative to reassess available systems of societal guidance and instruments of public accountability.

Finally, Nanette Newell, a molecular biologist who was project director for the second major OTA study on biotechnology, provides, as an important appendix to this volume, a discussion of the technique that created the controversy in the first place and the advances that biotechnology has provided us with in the nearly fifteen years since the first gene was successfully spliced.

These fourteen individuals have written separately but have also given us a unified view of these troublesome issues. One has seen an unknown creature from the front, others from the side, and still others from different perspectives. Together they may begin to construct a three-dimensional concept that will allow them to decide if it is a horse, an elephant, a squirrel, or something completely new. And when a fourth tells us how its skin feels, a fifth watches it running in a herd from a distance, and a sixth can only smell it, the true essence of the beast begins to emerge. If a seventh sees one being born, an eighth watches it eat, and a ninth sees it die, then we may begin to understand the total concept of the animal in both space and time. In this volume, we hope we have presented enough views of the life, times, and ultimate demise of the strange beast that is examined to perhaps recognize another if we see one, and predict its behavior. The next time, we may know enough to tame the wild creature before it runs amok again.

Every innovation brings both threat and promise, and our new powers of intervention in the depth of life are far from having run their course. The thesis of the planners of this book—if they have a single one—is that effects of major technological advances can be anticipated if guidance is sought in past experiences. This volume looks back in order to look forward. Not intended in any sense to rekindle controversies, nor to establish which of the gladiators in the gene-splicing wars were right or wrong, it emphasizes instead lessons to help us handle similar issues that surely lie before us.

RAYMOND A. ZILINSKAS
BURKE K. ZIMMERMAN

Editors' Note

Most of the papers included in this volume were originally prepared for a symposium that took place at the annual meeting of the American Association for the Advancement of Science in January 1982. Others contributions were invited shortly thereafter, when it was determined that the symposium would be published.

While the editors have tried to conform time references in the text to the expected date of publication, no attempt has been made to ask the contributors to update their discussions to take into account the relevant events, or the very significant technical advances, that have taken place during the past several years.

The collection of papers contained in this volume provide different perspectives and analyses of a significant period in the history of science and technology, from which certain general conclusions can be drawn and predictions made with regard to the future interactions between scientific advances and the larger society. In the fleeting period of ten years since the NIH guidelines were first issued (23 June 1976), the attention of the informed public has shifted from concerns about the safety of laboratory procedures to the wider question of their safe, beneficial and equitable applications. A discussion of these developments and their implications is appropriately left to the historians of the future.

The Gene-Splicing Wars

The evolution of the recombinant DNA controversy was governed by two distinct kinds of processes. On the one hand, there were the more or less predictable human responses—the expert practitioners in a new and esoteric branch of research called for self-restraint; others outside of this group saw events differently and, in many ways, misunderstood them. The fear of that which was not fully understood was a significant driving force in much of the contention that followed, and that fear was shared by all the major warring factions. But there were also a number of singular events, branch points at which certain actions determined which of several possible paths would be followed to the next phase. These paths were generally not predictable, yet their effects were of great significance. From our present perspective, it is possible to identify these points—they are few in number. The first was, of course, the Singer-Söll letter form the 1973 Gordon Conference. Later there were the doomsday article by Liebe Cavalieri in the *New York Times Magazine*, the public meeting of the NIH Director's Advisory Committee, and the unilateral decision of NIH to make the guidelines mandatory for NIH grantees. There was also that most significant early event, the Asilomar conference held in early 1975. Clifford Grobstein examines how this crucial event set the course of the policy discussions that comprised the core of the controversy for several years to come.

Asilomar and the Formation of Public Policy

CLIFFORD GROBSTEIN

The public-policy debate that was set off by laboratory recombination of DNA, a spectacular advance in genetic knowledge in 1972, has now lasted more than a decade. It is turning out to be almost as novel an exercise in public policy-making as it was a novel, scientific method. It is not novel in this century for technical advances to generate need for new public policy. In fact, repeated efforts to generate such policies have made the approach to resolving conflicts stemming from technological innovation itself the subject of controversy. Such approaches are evolving largely through trial, error, and the clash of powerful political pressures. Issues raised by recombinant techniques for genetic modification—like many others in the biomedical area—are especially emotive, substantively complex, and particularly sensitive by virtue of their high value content. The recombinant DNA controversy displayed all of these characteristics from the beginning. More recently, economic profit and the relationships among academic, corporate, and governmental roles in economic productivity have added to the complexity of the issue.

I shall focus here on the recombinant DNA debate as a public policy process. I visualize a broad policy "weather front" of which the issue of recombinant DNA and biohazard—as viewed in the mid-1970s—was only the first squall. Driving the broad frontal system is a capability for more penetrating and precise intervention in genetic systems of all organisms, potentially including the human species. Capability to alter genetic properties, and hence developmental processes, confers power to alter generation of organisms by *design*. Though such power is not entirely novel, the new increment is formidable and could have an unprecedented impact on human culture and the human species itself. Should this power be used? Can it *not* be used? If it is used, what and whose purposes should be served? By what mechanisms should purposes be sanctioned and by whom? How adequate are existing policy mechanisms for dealing with such issues?

In this perspective we may look back ten years and ask, what evaluation is to be made of the Asilomar conference of 1975? It clearly was a crucial event, both in opening the public phase of the recombinant DNA controversy and in shaping the fundamental concept of immediately subsequent policy. Was it, as some believe, a model of scientific initiative and social responsibility, or was it, as others believe, a disaster resulting from timid and fuzzy thinking? Whatever its success or failure in accomplishing its short-term objectives, did it trigger constructive or destructive longer-term consequences? In the context of what has happened since, should Asilomar have occurred?

Hindsight clearly indicates that whatever else it was conceived to be at the time, Asilomar was, in fact, the first step in a public policy process. In the view of its convenors it was a conference of scientists involved or likely to be involved in recombinant DNA research and with responsibility to think of the public welfare. These scientists would consider the issues raised and decide what should be done next. Its organizers were respected members of the involved group, and they largely determined the composition of the conference and its agenda. However, in addition to some 150 molecular biologists and geneticists (two-thirds from the United States), there were two small "external" groups, one of lawyers and one of journalists. The presence of these groups indicated the convenors' awareness that the matters to be discussed touched interests and issues beyond those of the immediate scientific group involved. The larger society was to be kept informed particularly since its legal and political processes might be invoked.

The conference concerned matters recognized to be of considerable uncertainty. Recombinant technology was expected to be a powerful new tool of great theoretical and practical potential. But of the many things that might be done with it, some could be conceived to have unanticipated and unfortunate consequences. No one knew for sure which or how many of these risks were real, but some seemed probable and serious enough to warrant deliberation and greater knowledge before undertaking them. The task of the conference was to devise rational and effective management of these risks while asuring continuance of the research.

It is important to recall that the conference was convened under the informal auspices of the National Academy of Sciences and with the encouragement and financial support of the National Institutes of Health (NIH). The Academy connection symbolized involvement of the broader scientific community. The NIH connection stemmed naturally from the fact that this agency was the major supporter of the research that led to the new technology. This, in turn, derived from public policy fashioned in the early 1950s assigning major responsibility for funding basic biomedical research to the federal government—operating through a chain of command from the President and Congress, through the Department of Health, Education, and Welfare (DHEW) to the Public Health Service and thence to the NIH. The scientists at Asilomar could see the nurturing, familiar, and friendly hand of NIH, but the arm and corpus of the federal government, with its responsibility for public policy, were only dimly perceived—and largely as a threatening image whose activation was to be avoided if at all possible. In fact, an advisory committee to the NIH director had already been constituted and was awaiting the conference results. This fact, together with the subsequent role of the Recombinant DNA Advisory Committee (RAC), makes it clear that Asilomar, private and primarily scientific as it may have seemed to the conferees, was a first step in consideration of public policy with respect to the new technology.

The precipitating motivation for the conference was risk, what came to be called biohazard. Could organisms genetically modified in the laboratory be dangerous if they escaped laboratory containment? In the most threatening scenario, could new sources of pestilence be created that would endanger the public health? The conferees were aware that other concerns than this kind of risk might be raised, for example, whether the already demonstrated exchanges of genetic information among widely diverse organisms constituted excessive manipulation of the natural order. However, it was agreed that such broader and even more difficult issues were not likely to be resolved in the scheduled three days of the conference and were not as immediate in possible consequence as biohazard. Moreover, they extended beyond the scientific competence and probably the capability for consensus of the participants. Therefore, the organizers decided to concentrate entirely on the risks of escape from laboratory control of genetically modified organisms.

The basic strategy adopted by the conference was to classify all conceivable recombinants as to their risk probability and to prescribe suitable risk-mitigating procedures appropriate to each. The prescriptions were to be administered as guidelines through the NIH, the major research supporter. The mechanism of administration was unspecified, but the term guidelines implied a voluntary system rather than a regulatory one.

What evaluation can be made of Asilomar as a device for initiating policy formation? Consider first its scientific aspects. It is clear, with hindsight, that the conference was too narrowly constituted (and too brief) to assess conceivable risks effectively. For example, molecular geneticists who made up the bulk of the conference had little expertise in microbiology, infectious diseases, epidemiology, pathology, sanitary engineering, public health, and other disciplines bearing on the human infectious disease risk questions raised. There is wide belief today that the limited expertise led to too casual acceptance of an excessive risk assessment. Moreover, the catalog of risks beyond the human also was limited by lack of expertise present in other important disciplines, for example, in plant genetics, physiology, and pathology.

How does Asilomar look in terms of its political aspects, that is, as a device for initiating a sound decision process where many interests (stakeholders) are involved? As seen subsequently by others who did not participate in Asilomar, it was a single-interest strategy session whose agenda was dominated by the particular concerns of the organizing interest group. To the degree that this was true, whether by intention or happenstance, Asilomar departed from open policy making in which all involved interests participate in agenda setting and have full access to the decision process. Indeed, this deficiency became a major and continuing issue as the controversy widened.

How successful was Asilomar in achieving the objectives set for it by its convenors? If the ojbectives were those earlier referred to, especially to devise rational and effective management of biohazards while assuring continuance of the research, the conference was highly successful. Research using recombinant techniques has suffered some irritating restrictions and delay but nonetheless has flourished. There has not been a single incident fitting the imagined risk scenarios in the ten years since Asilomar. The rationale the conference provided permitted both risk assessment and reduction of risk-containment measures as the level of perceived risk has declined. Today most research utilizing recombinant DNA technique proceeds without restriction. In terms of its short-term effects in relation to stated intentions, Asilomar was an unprecedented master stroke. Involved scientists recognized a problem, took an initiative directed to the public interest, devised an interim solution, and succeeded in defending it against significant opposition that would have imposed far greater restriction or, in the words of the most militant opposition, "shut it down."

What about long-term effects? What were the items excluded by the Asilomar emphasis on potential biohazard and what has happened to them? There are four main areas of subsequently identified potential concerns: (1) effects of scale-up resulting from industrial utilization, (2) the possibility of efforts to create harmful organisms deliberately, (3) environmental and ecological impacts of accidentally or deliberately released genetically exotic organisms and (4) applications to the human species. In effect, the Asilomar strategy was to relegate these items to secondary consideration, but in the long run, each may prove more consequential than the matter of biohazard taken only in the limited sense of an inadvertent threat to human health.

In this context a quirk of the Asilomar strategy is highly significant. As noted earlier, the rationate for NIH involvement in the Asilomar process was clear and logical to biomedical scientists. NIH was both the major supporter of relevant research and sensitive to the views of the involved scientists. If government were to be in any way involved, NIH was the most acceptable agency. However, as a public agency NIH had a defined mission and could not avoid public accountability. Moreover, it was answerable to the Secretary of DHEW. These legal and political realities became clear as soon as the Asilomar recommendations began to be turned into formal governmental action. The Asilomar strategy thus was inevitably tilted toward at least quasi-regulatory federal intervention, using an agency that was not regulatory in mission or experience and hence had to improvise its administrative course.

Nor did the mission of the NIH cover all of the agenda items, Asilomar excluded. The NIH mission was reasonably concordant with the biohazard issue because it had, as former NIH Director Fredrickson has said, ample relevant expertise to consult among its own staff and that of such sister agencies as the Center for Disease Control. But environmental and ecological considerations, and agricultural applications in particular, were beyond both its expertise and its mission. Thus, assignment of the matter of recombinant DNA policy to NIH leadership solidified the restricted agenda defined at Asilomar. Despite the later creation of a NIH-led interagency committee for federal coordination, a forum for concerted deliberation on the excluded Asilomar agenda never came into existence. So the public policy debate and, to some degree, the public impression of recombinant technology remained fixed on worst-case scenarios symbolized by the Andromeda strain. As that perceived threat receded, it was almost immediately replaced by the Midas touch, the risks of venture capital, and the problems of university-corporate-governmental interaction.

Was Asilomar therefore a mistake? My own view is that it is not useful for hindsight to try to diagnose error in judgment in order to assign blame. Hindsight, like foresight, has limitations. Many who lived through Asilomar, and became battle-weary thereafter, came to regret that it ever occurred. Speaking personally again, I see this view as understandable but misguided. The public image of science was humanized and thereby enhanced. Asilomar is not interpreted generally as a brilliant conspiracy but as a conscientous effort by well-intentioned people who did not know what was best to do but did their best despite conflicting personal interests and inclinations. Asilomar also protected the extension of knowledge for a smaller price than might otherwise have been paid.

Mistake, however, is not the issue. That hindsight could yield better options than Asilomar is certainly a possibility. Some charge that the conference was overreaction to uncertainty, that not even the wider community of scientists should have been alerted or consulted. That course was followed, perforce, some thirty years earlier in the atomic bomb decision, when national emergency and military security dominated the process. In the early 1970s, not many supported a similar option, especially if they saw a parallel. However, some participants at Asilomar saw a "trap" in inviting wider governmental involvement and would have preferred to have quietly resolved uncertainty within the group of scientists who were best informed. This "insider" approach is a true option to the uncertainties and vicissitudes of public policy making and continues to have its proponents. It is, indeed, a basic, continuing issue, and in policy terms the fundamental one. For scientists, who find their activities of increasing social consequence, the issue is how to manage their side of the implicit moral contract with society to do no harm while preserving sufficient independence to promote vigorous inquiry. For nonscientists, whose values and ways of life are increasingly pressured by the seemingly uncontrollable thrust of science and science-derived technology, the question is how to derive benefits but still have a voice in assessing risk and choosing among options often presented in incomprehensible terms and contexts.

Thus, the most consequential criticism of Asilomar is that it was public decision making without appropriate public access, in extreme terms, a covert exercise of power by a special interest group. Although the matter is complex—and the charge in some ways ironic—it has sufficient credibility to deserve careful consideration. Certainly it is true that Asilomar set in motion a policy pattern that was preserved against subsequent bitter protest, both within and outside the scientific community. It is also true that Asilomar itself provided only limited public, or even general scientific, access to its decision making. It is clear, however, that the translation of the Asilomar recommendations into promulgated guidelines had public access, though whether it was sufficient is debated. Certainly efforts were made by the NIH director to comply with federal requirements for broader access and wider perspectives. Certainly, too, although the Asilomar pattern was preserved into the guidelines, significant changes were made in response to various pressures. Moreover, later press coverage promoted the issue to the status of major news, and congressional hearings voluminously augmented the public record. Nonetheless, if Asilomar is looked at not as a scientific conference but as a first step in public policy formation, then in this controversy of some consequence an initial and ultimately decisive step was excessively dominated by a single interest to the exclusion of all others.

Perhaps the best statement is that the Asilomar conference was ambiguous in its execution and possibly in its conception. It began as a scientists' strategy conference and became a step in public-policy formation. In process it was the former, in consequence it was the latter. Such ambiguity will not give credibility it was the latter. Such ambiguity will not give credibility to science or scientists in an open and rigorous public-policy area in which contending interests insist on their prerogatives.

The point is not without current relevance. The United States for the past half century has recognized a public and a private sector and made choices in a persistent tension between the two. The proper size and role of each is undergoing reevaluation at the moment. Recombinant DNA technology has become a cutting edge in a rapidly expanding biotechnology. The effect of commercialization of so recently an essentially completely academic and governmentally supported activity is being scrutinized, largely in terms of the implications of the profit motive and appropriate allocation of expected economic returns. Less attention has been paid to the fact that a sensitive, value-laden technology is becoming increasingly the province of corporate board rooms and therefore less accessible to public scrutiny and decision-making. This accompanies a trend toward proprietary operation of health care systems, likewise less accessible to public scrutiny and decision making. This "insider" problem will heighten the debate over public participation in scientific decision making that already has figured prominently in the recombinant DNA controversy.

The matter is too complex for a full discussion in this brief chapter. But it should be noted that the Asilomar recommendations made no reference, and the conference gave no consideration, to nonscientific representation in scientific decision making. Nonetheless, pressure on the NIH led to nonscientific representation not only on the advisory body to the NIH director but on all local institutional committees reviewing individual projects. This was one manifestation of the assertion that the consequences of scientific and technological advance for the general public have now become so massive that public decision making must have not only access to but influence on scientific and technological directions. Whether this view will gain or lose adherents only time will tell. Nonetheless, there is something to be learned from tracing the course of this thrust, considering its rationale and justification and, particularly, conceiving ways to enlist its legitimate aspirations to fortify and not endanger the rigor and independence of scientific activity.

What summary evaluation can be made of Asilomar in the light of hindsight on its policy-forming role? First, it was an inspired improvisation to meet difficult and seemingly unprecedented circumstances. Neither participants nor the scientific community should regret or apologize for either its objectives or its accomplishments. On the other hand, as a policy-making mechanism it had defects to be frankly acknowledged. The key defect was the failure to recognize that an issue large enough to halt, even temporarily and partially, the normal process of science requires consideration by more than the group of scientists involved. The sponsorship of the National Academy of Sciences should have been more than symbolic and NAS should have ensured that the scientific community fully understood the issues, participated in the analysis, and endorsed the recommendations. If the Academy is not the appropriate body for this role, the scientific community needs to decide what body is. The issue of inadvertent biohazard may have been put to rest, but the agenda items excluded at Asilomar have not. That defect of Asilomar should not be repeated.

The issue faced at Asilomar, and certainly the excluded issues, require attention beyond the scientific community. The presence of small groups of lawyers and journalists at Asilomar acknowledged this but did not meet the requirement. In an open society public decision making requires access by all interests who will later be called upon to support the decision. Private caucuses have their place, but they must not be confused with public-decision arenas. Nor should private caucuses be allowed to set the public agenda. This subverts democratic processes of policy making. The scientific community should be seeking additional ways to cooperate with the larger community in making policy with respect to issues of high scientific and technological content. This is not best accomplished by self-selected "inside groups," whether of experts, or of leaders of particular interests, or of corporate boards.

The problem is not an easy one. But in the course of the recombinant DNA controversy improvisations other than Asilomar were successfully used to broaden the base of decision making. This happened not only in Congress and in federal agencies but in such local communities as Cambridge, Ann Arbor, Princeton, and San Diego. These innovations need careful consideration as indicators of the kind of mechanisms that ought to be strengthened to meet the oncoming policy weather front induced by our heightened capability for genetic and developmental intervention. These cut too close to the human core to be handled insensitively. In such an area science and technology must stay close to broader public purpose. This will require new and wider forms of communication and new accessories to public decision making. Of this, as of recombinant DNA, we have so far seen only the beginning.

A committee of the National Academy of Sciences, chaired by Paul Berg of Stanford, published a letter in *Science* on 26 July 1974 (p. 303), in which it recommended future steps to deal with policy concerning genetic manipulation in the United States. One of these recommendations, for an international conference, led to the landmark Asilomar conference. Another asked for the voluntary deferral of experiments that might introduce the genes for antibiotic resistance or bacterial toxin formation into strains which do not contain such genes, or which would link the DNA from tumor-inducing or other animal viruses to bacterial plasmids, and called for prudence in linking animal DNA to that of lower forms of life.

The other recommendations set in motion the events which were to establish the National Institutes of Health's central role in the controversy in the years ahead. Specifically, "the director of the National Institutes of Health is requested to give immediate consideration to establishing an advisory committee charged with (I) overseeing an experimental program to evaluate the potential biological and ecological hazards of the above types of recombinant DNA molecules; (II) developing procedures which will minimize the spread of such molecules within human and other populations; and (III) devising guidelines to be followed by Investigators working with potentially hazardous recombinant DNA molecules." Thus was born the Recombinant DNA Molecule Program Advisory Committee and, in time, the NIH Guidelines for Research involving Recombinant DBA Molecules.

Donald Fredrickson returned to the National Institutes of Health as its director in the summer of 1975, after a year as president of the Institute of Medicine, just as the deliberations over gene splicing were getting interesting. For the next six years, it was his fate—or privilege—to bear the responsibility for the content of the NIH guidelines and all of their revisions. This, and all of the attendant activities required to enable him to bear this burden, took more of his time during his tenure as director than any other single issue with which he had to contend. Here is his account of those historic years.

2

The Recombinant DNA Controversy: The NIH Viewpoint

DONALD S. FREDRICKSON, M.D.

Revelation from human history takes time. And, as the gospels have shown us, both strong belief and the opportunity to work it out as a community also are helpful. This volume should help all of us weigh what we respectively believed and did during the recombinant DNA controversy, and judge the appropriateness of these actions with hindsight. A second and more important purpose is to consider not so much what we did as what we are to do in the future. This includes the future uses of recombinant DNA technology as well as the manner in which we should handle other controversies of a similar kind.

It is my assignment in this chapter to cast reflections from the many-surfaced mirror of the federal government. What did it do? What did it seem to do right—or wrong—and why? In the limited space available, I would like to state my views on some things the government did right, and where it might have gone wrong. Of course, I will have some biases, and this is the first occasion since leaving the NIH directorship that I have had to air them.

NIH as Lead Agency

In my opinion, one of the good moves of the federal government was that it let the National Institutes of Health carry the principal federal responsibility in the controversy.[1]

In 1977, one famous scientist wrote, "I consider it a true calamity that the agency dispensing nearly all the federal funds available for biological and biochemical research, NIH, has become a party in the debate."[2] Elsewhere, he opined, "the National Institutes of Health have permitted themselves to be dragged into a controversy with which they should not have had anything to do," and, "our time is cursed with the necessity for feeble men, masquerading as experts, to make enormously far-reaching decisions."[3]

Another critic wrote in the same period, ". . . an ethical conflict of interest arises when it [NIH] is entrusted to set up guidelines to regulate the very research it is committed to promoting."[4]

I agree with the appearance of a conflict of interest. It was unavoidable. It was bothersome all the way. One of the most important lessons to be learned about controversy over use of high technologies, however, is the absolute requirement for expert opinion. The most informed experts very often will include those using or promoting the technology, and so there will be an appearance of conflict of interest. The art of solving this kind of problem lies in the manner in which one joins the experts with the other parties at interest.

The NIH had at least five distinct advantages that made it the agency of choice for establishing guidelines and providing the focus for federal activities concerning recombinant DNA:

1. NIH originally had been asked by the scientists involved to help. In the often-cited letter to *Science* of 26 June 1974 to the director of NIH, eleven scientists acting for the Assembly of Life Sciences of the National Research Council and including some of the key molecular biologists who were to attend the Asilomar Conference requested NIH to start a program to evaluate the biological and ecological hazards, to consider procedures to minimize the spread of recombinant molecules, and to devise guidelines for investigators.[5] NIH therefore had the confidence of the scientific community most directly involved.

2. NIH was funding far more of the research involving recombinant DNA technology than was supported by any other source. It therefore had the "clout" to enforce adherence to guidelines if that became necessary.

3. The most important advantage of NIH was that it was there—staffed, integrated into government, and ready to go. The recombinant DNA controversy needed to be dealt with "on-line" and within the existing framework of government. It can be a grievous error to assume that dilemmas involving profound questions about science should automatically require new, untested solutions. In the first three to four years of the controversy, I believe it would have meant chaos had this problem been handed to a commission made up of busy citizens, no matter how distinguished, who could give it only part-time attention.

4. Among all federal science agencies, NIH had a unique feature

whose essentiality in such a controversy was not immediately recognized. This feature was the great size and quality of its intramural research program. The presence on the NIH campus of many experts in the techniques in question—scientific peers of those in the extramural community—made it possible for NIH to weather the storms that blew up around the speculative hazards and the threats to scientific inquiry inherent in the crisis. From among the staff I quickly assembled the NIH "Kitchen-RAC," which counseled and crafted solutions to endless problems as this complex and lengthy transaction proceeded. As director, I spent a third to a half of my time on recombinant DNA in 1976 through 1978. This was but a small fraction of the total NIH person-hour expenditure. I would have wasted those hours of mine but for the dedicated and talented scientific and administrative NIH people always at hand. Praise of these persons is one of the neglected choruses in the recombinant DNA epic.[6]

5. NIH was a science agency without formal regulatory experience or authority. Had it been an agency so endowed, such as, for example, the Food and Drug Administration (FDA), the Centers for Disease Control (CDC), or the Environmental Protection Agency (EPA), it could not have kept the setting of standards and their revision out of the chilling grip of conventional regulation. This would have involved the tedious, stepwise processes specified under the Federal Administrative Procedures Act. Every decision arising from the stream of new knowledge would have had to be diverted into the *Federal Register* for publication and comment. The risk of slowing the evaluation of the science into a sludge of unfinished experiments and unsatisfied hypotheses was too great. Rather, with the concurrence of superiors in the Department of HEW, I decided we should take advantage of the previously established practice of NIH directors to impose certain conditions upon scientists as guidelines. To this we would add the specifications of the Federal Advisory Committee Act protecting public access to the decision-making process. We would refrain from the formal *Notice of Proposed Rulemaking* which would have placed us in lock-step with the regulatory process. To be sure, well-informed critics such as former FDA General Counsel Peter Hutt took the NIH to task for its unseemly amateur performance as a regulator,[7] but we were determined to walk the narrow edge of the abyss until a special procedure for evolving the guidelines, consonant with the pace and scale required to synchronize experiments in scores of laboratories, could be established. As it turned out, this took two years of departmental negotiations and public hearings.[8]

Ecumenical Executive Agencies

Once harnessed, all the government executive agencies, research and regulatory, worked harmoniously.

In the beginning, there were at least four government agencies supporting recombinant DNA research (National Science Foundation, Department of Agriculture, Veterans Administration, and NIH). These agencies, it quickly became apparent, would need to agree upon a single set of standards. Moreover, there were a half-dozen, including FDA, EPA, CDC, the Occupational Safety and Health Administration (OSHA), the National Institute for Occupational Safety and Health (NIOSH), and the Federal Transportation Agency, which believed their authorities permitted them some regulatory authority over the products of such research, and perhaps even the laboratory experiments. As soon as we had guidelines ready for promulgation, I obtained the agreement of David Mathews, then Secretary of the Department of Health, Education, and Welfare (DHEW), to urge President Ford to convene an interagency committee. It was to include all of the above federal agencies and others which had special interests in the problems, including the Departments of Justice, Defense, and Commerce, the Council on Environmental Quality, and the Office of Science and Technology Policy (OSTP).

Months went by with no issuance of a presidential order. Disagreements among agencies over jurisdiction increased. Finally, Senators Kennedy and Javits issued an open letter demanding that the Interagency Committee be formed. The White House acted and the committee, which I chaired, got down to business in November 1976. Fairly quickly it achieved three objectives:

- All the research agencies (despite some sacrifice in autonomy) agreed voluntarily to adhere to one set of standards and to support a single locus of interpretation (NIH).
- All the regulatory agencies submitted to a common examination of their enabling statutes and agreed that none had clear authority to regulate, except in limited areas.
- The committee concluded that a new law would be both required and desirable to assure that the NIH guidelines were followed in all similar research in the private sector.

So a prescription for a model statute was developed. It included preemption of all other standards by federal ones and a sunset clause. This was forwarded to Joseph A. Califano, Jr., then Secretary of DHEW. His general counsel drafted a bill to be sent by the administration to Congress for action to create the new law. When the draft was sent through the Office of Management and Budget for clearance, last-minute anxiety on the part of one federal agency about the Secretary of DHEW having authority over the experiments conducted by its scientists threatened to send us all back to the drawing board. The government bill survived to be introduced. It was quickly lost from sight, however, as I will relate later.

All the government agencies also were given liaison membership in the Recombinant DNA Advisory Committee (RAC), where decision making under the guidelines proceeded. An Industrial Use Subcommittee, under the chairmanship of Gilbert Omenn of OSTP, was formed within the Interagency Committee to consider concerns raised by NIOSH and OSHA about the risks of industrial scale-up for recombinant technology.

There was considerable concern over turf within the federal bureaucracy, recombinant DNA being a matter of high press and public interest. The mandates of regulatory agencies do not allow them any rest if any possibility exists of their being considered derelict in a responsibility. The Interagency Committee allowed anxiety to be relieved by augmented communication and frequent discussions. In addition to maintaining a desirable amount of ecumenical spirit, this committee also had the virtue of being in place and ready for immediate convocation in the event that one of the hypothetical hazards did materialize and national resources needed to be mobilized and coordinated.

Other Struggles for Jurisdiction

The controversy over the use of recombinant DNA technology was not limited to Washington. Indeed, in late 1976 and 1977, the state of New York and several communities such as Cambridge, Massachusetts, and Ann Arbor, Michigan, were the sites of vigorous debates. From them arose pressure for enactment of new local or state laws, threatening a Balkanization of the scientific effort through differing regulations for conduct of laboratory experimentation. Such situations are rare in modern history and pose an unusual threat of disruption of scientific activity. This was the background against which the federal Interagency Committee conducted its work and the RAC proceeded to consider how to revise the original NIH guidelines that set the rules for all federal supported laboratories after 23 June 1976. It was the extension of these rules to all laboratories, whether receiving federal support or not, that partly maintained the controversy. Disagreement also existed as to whether these guidelines were too strong or too weak. For many, the interests lay in the procedures the scientists were ready to follow in revising the rules, particularly if they intended to relax them. The controversy became hotter when the Ninety-fifth Congress convened in January 1977.

Congressional Caution

The Congress, despite the introduction of more than a dozen bills and intensive hearings on the subject, eventually refrained from enacting a statute to control laboratory experimentation with recombinant DNA.

The activities of the Congress relative to recombinant DNA merit a more thorough and thoughtful analysis than is possible in this chapter. One would like to explore more carefully the various motives of the legislators and their staffs that impelled them to propose legislation. Needing detailed description, too, are the hazards of drafting statutes to restrict scientific freedom in a single, highly technical area. The play of forces that ultimately led to a stalemate, no bill actually coming to a vote in either house, is a theme for several essays. There were not only conflicts over passage of new legislation but also over interpretation of old laws to achieve the same purposes. Of particular relevance here were the attempts to make Section 361 of the Public Health Service Act (42 U.S.C. 264) the basis for nationwide regulation of recombinant DNA research. This little-used section permits the Surgeon General to take steps he deems necessary "to prevent the introduction or spread of communicable disease," a potential hazard of the use of recombinant DNA technology. For this there was a notable absence of proof. Many members of Congress, as well as the secretary and the general counsel of DHEW, and the Surgeon General joined NIH in opposing use of Section 361. There was a general feeling that if Congress wished to regulate laboratory experiments in biology, the members should stand up and be counted.

A rereading of the bills submitted reveals, amid the boilerplate, some intimate glimpses of tensions experienced by the congressional sponsors. Some of the "Whereas's" were followed by dire predictions, others by acknowledgment of wondrous benefits to accompany any hazard. The kinds of fines and penalties to be assayed showed how clumsy and unrealistic are the provisions of statute for governing this kind of human activity.

The most provocative piece of legislation proposed was S. 1217 (Amended), introduced by Senator Edward Kennedy in July 1977. The initial bill, which he had submitted in April, was the administration's minimal proposal prescribed by the Interagency Committee.

I supported this original "administration bill." It provided for federal preemption of any local regulations. Such preemption is always a controversial matter. It is consonant with the essential universality of science, however, and—something more practical in the recombinant DNA affair—it was in keeping with the absence of imminent danger to any particular community. The sunset provision of the administration bill also was a comfort. Any legislation over so mobile an activity as scientific research should have limited life expectancy. As in the other bills, the penalty clauses were harsh and foreign to scientific research; but they were a bearable price, if we had arrived at the need for federal legislation in order to allow experimentation to proceed.

But S. 1217 (Amended) also contained a new Title XVIII establishing a National Recombinant DNA Safety Regulatory Commission. The body would be serviced by DHEW but not clearly answerable to the Secretary. Of its eleven members, six were never to have engaged in molecular biology. Thus, a new administrative creation would be established to supply the consensual requirements, the majority of participants to be inexpert in the subject matter. The commission would set the rules (as neatly promulgated regulations), license laboratories, and monitor compliance. Any extra time left over during the periodic visits of its members to Washington was consigned to a thorough analysis of "all the basic, ethical, and scientific issues involved."

The recombinant DNA controversy excited the natural tendency to assume that dilemma involving profound new questions should automatically be exposed to new, untested mechanisms for coping with them. Fortunately, the maturity of our political system—or the stubborness of its traditions—forced this complex problem to be engaged first within the existing framework of government. The recombinant DNA issue would have been gravely confounded by immediate convocation of one or another of the ad hoc commissions contained in numerous bills and articles stimulated by the controversy. One proposed that the Vice President chair the proceedings; another invoked a science court. One quickly learns in public service that there is no such thing as "immediate" chartering, staffing, and convening of a commission for any purpose, let alone preparing the first attempts at statutory regulation of complex laboratory experiments in biology.

One evening during the legislative furor over recombinant DNA, I was summoned to the bedside of Congressman Olin Teague at the Naval Medical Center. "Tiger" Teague was chairman of the House Science and Technology Committee. He sought a full explanation of the new biotechnology and an opinion about possible effects of pending legislation upon the progress of science. He questioned me for over an hour and listened carefully to the answers. Later, I heard how Teague had made sure that the House bills to regulate recombinant DNA by statute were sequentially referred to his committee for a long and thorough hearing. In so doing, he provided a way for tempers to cool and protected us all from hasty passage of laws that would have been injurious. When NIH and DHEW devised ways for industrial and other private-sector laboratories to comply voluntarily with the NIH guidelines, the pressure for legislation was nearly gone, and has not since been revived.

Imaginative Structures

Government was not devoid of imagination in creating new administrative structures to permit the public a role in the recombinant DNA controversy. The "second generation" RAC was perhaps the outstanding case in point. Reorganization of this committee was a concession demanded by Secretary Califano for permission to release the first major revision of the original guidelines in December 1978. The original RAC, formed in 1975, had contained only scientists, nearly all of them molecular biologists. A political scientist was then added and somewhat later an ethicist came aboard. We initially employed the NIH Director's Advisory Committee, suitably augmented with a broad selection of scientists and laymen, as a second tier for review. It was the traditional organization selected by the Congress for the greatly expanded public support of basic research that commenced in the early 1950s.[9] The system is based on initial peer review (study section) followed by oversight of a group including laymen (advisory council) and has proved admirable for determining the allocation of resources for research. It is not effective for supervision of technical guidelines requiring continuous and rapid evolution.

DNA technology was exceedingly complex material, heavy going for laymen, but also for scientists from other disciplines. Procurement of advice and approvals in two stages created confusion and added intolerable delays. Hence, the RAC was changed to collapse review into one group. Its membership was composed of one-third molecular biologists, one-third scientists who were experts in genetics, microbiology, and other fields directly applicable to recombinant work, and one-third experienced in related matters such as public health, law, consumer affairs, or public policy. I have observed, particularly in the technical consensus exercises we established at NIH in the same period,[10] that when the nonexpert is not able to comprehend much of the detail, his public-policy role may better be performed in the midst of the experts. Here, at least, the lay person can observe the experts to see if they appear to be listening to each other and paying some attention to the evidence.

The new RAC, like the first one, remained advisory to the NIH director, who had the responsibility and authority for revision of the guidelines. Some of the scientists were alarmed when the new RAC was put into position,[11] but their fears of a shift of governance from the scientific to a political sphere did not materialize. I believe the new RAC was one of the most useful cultural innovations—for combining expert and nonexpert opinions about science—to emerge from the recombinant DNA controversy. Its success has been due to careful selection of members, to their generally enlightened individual performances, and above all, to the guidance of the chairmen. These have been Jane Setlow, a molecular biologist, and Ray Thornton, a lawyer and university president and once the congressman who conducted the Science and Technology Committee hearings on recombinant DNA fostered by "Tiger" Teague.

International Affairs

We were careful that NIH should not confuse its predominant place in the world of biological research with a mandate to determine the regulations that would govern recombinant DNA research in the rest of the world. The sovereignty of each nation over its scientists was not a debatable issue. The role of the United States was both a delicate and influential one. Initially, we considered it likely that there would be less controversy in most other nations and that conditions outside the United States might well favor migration of our scientists to areas more congenial to the experimentation. It was certainly true that the extremes of reaction were observed in America and that in some countries there was little or no concern. Nearly all of the countries with advanced science and technology did adopt rules, however. The United States, Canada, and the United Kingdom were the first to have explicit guidelines. The major counterpart to the RAC proved to be the British Genetic Manipulation Advisory Group (GMAG), which set out to construct rules for the United Kingdom following the Ashby report (see Chapter 4). The NIH guidelines were the first of numerous different national rules to be released. Care was taken to distribute them abroad. More than forty countries were sent the guidelines by diplomatic pouch on the day of release in 1976, accompanied by a mission alert from the office of then Secretary of State Henry Kissinger.

There was continuous follow up. Among my papers are special travel diaries entitled "The Recombinant Odyssey." They summarize my numerous visits to scientists and officials of countries which included, among others, Great Britain, Germany, Canada, Holland, Switzerland, the People's Republic of China, Japan, France, Sweden, Finland, Italy, and the Soviet Union, as well as the European Economic Community in Brussels, to explain what NIH was doing and to learn how the other nations were attempting to regulate the research. I recall paying an early visit to Munich to see the new chief executive of the European Science Foundation (ESF). The late Franz Schneider informed me that he was sure the ESF would adopt the U.K. rules. When I crossed the city to Professor Feodor Lynen's Max Planck Institute, however, I found a scientist busy translating the NIH guidelines into German.

After my first visit to the British Medical Research Council and to GMAG, I realized that we were likely to achieve a kind of parity with the United Kingdom on containment rules despite a completely different mode for achieving them. The British proceeded to develop common law, case by case. The United States specified detailed rules in a veritable Napoleanic Code. The British met in closed rooms, protected by the Official Secrets Act. The United States opened the doors to everyone and compiled a massive record of the proceedings.

We worked hard to assure conditions for maximum communications and consensuality among all the users of the "new biology." Officials of the European Medical Research Council, the European Community, the European Molecular Biology Organization, and the Committee on Genetics of the International Council of Scientific Unions were often present at the sessions of the RAC and the Director's Advisory Committee, which were always open meetings. One of two major meetings which the NIH sponsored in 1977 to help clarify scientific knowledge on which the guidelines were based was held in Ascot, England, so that British and continental scientists could more easily attend. The private sector, including early industrial users of the technology, and public interest groups much concerned about the activities of the former, were also tied into the loop of communications.

In fact, the NIH Office of Recombinant DNA Activities, headed by William Gartland, became the primary communications center in the world concerning recombinant DNA during the late 1970s. We also persuaded the Office of Management and Budget to let us start a new journal, the *Recombinant DNA Technical Bulletin*, to carry the actions of RAC and scientific communications around the world. The safety manual and other advisories compiled under Director of Research Safety Emmett Barkley's direction, was another of the many technical aids devised to help standardize certain practices here and abroad. The course of GMAG in Britain and the rules eventually adopted by other countries are described by Keith Gibson in Chapter 4. Most nations have adopted basically the NIH guidelines, which have remained commensurate with those of the United Kingdom.

Major differences in national standards would have precipitated a chaotic situation throughout the world—as would different regulations in the municipalities or states in the United States. Indeed, had guidelines of grossly uneven character sprung up within countries, or among them, the new biotechnology industry based on recombinant DNA methods conceivably would have risen in the Third World, or on ships "beyond the twelve-mile limit" in the fashion of gambling casinos.

OTHER STEPS

There were many other decisions and actions for the government to take about the uses of recombinant DNA technology. Each agency decision had to be shaped so that it could make its way past the checks and balances built into the federal process. For example, there was the decision to be made about NIH-DHEW policy on patenting inventions derived from the use of genetic recombinants. Opinions were solicited and a decision made acceptable to the Secretary. Shortly thereafter, I found myself before Senator Gaylord Nelson of the Committe on Labor and Human Resources, who had strong views about patenting anything discovered through use of public funds. Later revisions have considerably liberalized the policy he and I were discussing.

There was the decision to be made about how to return the authority for using the guidelines to the institutions where I felt the responsibility belonged. The composition and function of the institutional biosafety committees (IBCs) was another important exercise in political compromise. One of the most signifiant moves was the determination of how the RAC might review voluntary submissions of proprietary data from private sector laboratories interested in scale-ups. The RAC's eventual willingness to do this removed the last powerful thrust for legislation. The guidelines no longer require such examination. While they did, it was necessary to persuade the RAC regularly to continue this service, for it was an exercise which annoyed many of the members.

It will be impossible for some of us to forget the problems engendered by compliance with the National Environmental Policy Act. The first Environmental Impact Statement on hypothetical risks of laboratory experimentation became a nightmare before it was accepted. Yet it proved invaluable in opposing the injunctions against experiments that were sought in the federal district courts. The record maintained by NIH from the inception of its role contains more history of the roles pressed upon the government and the manner in which they were played.[12]

The Future of the RAC

Although we have learned that the probabilities of harmful creation from using recombinant DNA technology are much less than some believed in 1975, no one can assign a zero probability to harmful effects now. No one should pretend we have absolutely no further need for community guidance or for continuous evaluation of such powerful technology. It is, however, reasonable to ask whether the risks have not narrowed to those already assigned to the common vectors and hosts, so that special containment precautions for gene engineering may now be wasteful. The problems that remain to be dealt with—such as release of recombinant organisms or plants into the environment, evaluation of the numerous products of recombinant genes, or the effecting of changes in the human genome by new techniques—are still there. But so are agencies and other institutions to cope with the regulatory aspects of most forseeable problems. If some of the federal regulatory forces appear to be weaker than in 1975, it would be very unwise to compensate for this by forcing NIH to assume regulatory roles it has justifiably resisted for so long.

There is a continued need for full communication and critique in the use of recombinant DNA technology. It is the key expression of the continuity or universality of science. One would not want to abandon the network of Institutional Biosafety Committees and the RAC, which lies at the center, until they clearly no longer serve a useful function.

It is my view that we are not finished with practical scientific questions and ethical issues related to gene splicing. These will not be the stuff of conventional regulation, and they will require a proper place or places for the human community to debate and resolve them.

As the rules for conventional laboratory experiments inevitably slide toward the status of guidance, and as broader policy problems replace detailed analysis of experimental protocols on the menu for its consideration, we should think about construction of a "third generation" Recombinant Advisory Committee.

I suggest it might have these features:

- Be designed to fill the combination tasks of the present RAC and Interagency Committees
- Be responsible to a cabinet officer, the most appropriate still being the Secretary of Health and Human Services
- Continue to be serviced by NIH, but with broader contributions from other agencies so that the collective aspect of the future enterprise will be stressed and facilitated
- Continue to have a distinguished chairman, from the nongovernment sector
- Continue to have a majority of expert scientists among its members, but a total composition tailored to reflect the problems anticipated in the next few years

Post-Game Critique

I have described a number of things that the federal government did during the recombinant DNA controversy. Having been at or near the center of those actions, I am not the one to judge each step or to assign a mark for the overall performance.

Such an assessment should be undertaken, however, for high science will confront big government again with similar dilemmas. We need a clear understanding of what was done and why. The design of the federal government is such that the public interest in technologies can be served without impairing the effectiveness of the scientific endeavor. This was the major civics lesson to emerge from the recombinant DNA controversy. It was difficult, however, to maintain the proper balances, and one should not assume that the system will never fail.

Notes

1. The National Institutes of Health (NIH) is an agency in the Public Health Service which, in turn, is part of the U.S. Department of Health and Human Services (until 1977, the Department of Health, Education, and Welfare). NIH is responsible for more than half of the federal support to the universities for scientific research and development. In addition, NIH has an "intramural" research program, located principally in Bethesda, Maryland, which is the largest biomedical research institution in the world.

2. Erwin Chargaff, "Uncertainties Great, Is the Gain Worth the Risk?" *Chemical and Engineering News*, 30 May 1977, pp. 32–35.

3. Erwin Chargaff, "On the Danger of Genetic Meddling," *Science* 192 (4 June 1976): 938–40.

4. Francine Simring, "The Double Helix of Self-Interest," *The Sciences*, May/June 1977, pp. 10–27.

5. Paul Berg, David Baltimore, Herbert W. Boyer, Stanley N. Cohen, Ronald W. Davis, David S. Hogness, Daniel Nathans, Richard Roblin, James D. Watson, Sherman Weissman, and Norton Zinder, "Potential Biohazards of Recombinant DNA Molecules," letter to the editor, *Science* 185 (26 July 1974): 303.

6. The daily menu for the "Kitchen-RAC" (named after the parent NIH Recombinant DNA Advisory Committee) was usually prepared by Joseph Perpich, associate director for program planning and evaluation, and Bernard Talbot, special assistant for intramural affairs. Perpich's combined medical and law degrees, plus a clerkship with Judge David Bazelon and time on the staff of Senator Edward Kennedy, enabled him to provide me with invaluable advice on meeting both legal responsibilities and political objectives. His specialty training in psychiatry also came in handy. Both an M.D. and a Ph.D., Talbot was the perfect antidote for pejorative views on productivity of government employees. His Stakhanovite work habits enable him to produce mighty drafts and redrafts of revisions of the highly technical guidelines in response to endless commentary and pressure for alterations. Other invaluable contributors were Emmett Barkley, director of the office of research safety; William Carrigan, editor of the NIH papers on recombinant DNA; William Gartland, director of the office of recombinant DNA affairs at NIH; Susan Gottesman, a scientist in the laboratory of molecular biology in the National Cancer Institute and a member of the NIH Recombinant Advisory Committee; Joseph Hernandez, an attorney and member of our division of legislative analysis; Malcolm Martin, a virologist and molecular biologist from the National Institute of Allergy and Infectious Diseases (NIAID); Richard J. Riseberg, NIH's legal advisor, whom I once called "a double agent with cover blown from the start," because he was officially in the DHEW General Counsel's office; the late Wallace Rowe, a famous virologist, member of the RAC, and laboratory chief at NIAID; Betty Shelton, whose staff had a prodigious capacity for production of copy; and Maxine Singer, a Cancer Institute molecular biologist who had been in on the recombinant DNA controversy from the start, and whose contributions toward its resolution were both legion and indispensable. Burke Zimmerman, who joined us later on, brought with him the valuable perspectives of the environmental groups and of the congressional staffs.

7. Peter Hutt, letter to Donald S. Fredrickson, 3 March 1978, NIH Papers, Appendix A, pp. 239–56.

8. Donald S. Fredrickson, "A History of the Recombinant DNA Guidelines in the United States," in *Recombinant DNA and Genetic Experimentation*, Joan Morgan and W. J. Whelan, eds. (Oxford and New York: Pergamon Press, 1979), pp. 151–56.

9. Don K. Price, "Endless Frontier or Bureaucratic Morass?" *Daedalus* 107, no. 2 (Spring 1978): 75–92.

10. Donald S. Fredrickson, "Seeking Technical Consensus on Medical Interventions," *Clinical Research* 26 (1978): 116.
11. Maxine Singer, "Spectacular Science and Ponderous Process," editorial, *Science* 203 (5 January 1979): 9.
12. Office of the Director, National Institutes of Health, *Recombinant DNA Research; Documents Relating to "NIH Guidelines for Research Involving Recombinant DNA Molecules"*: Volume 1, February 1975–June 1976, DHEW Publication No. (NIH) 76–1138, August 1976, 602 pp. Volume 2, June 1976–November 1977, DHEW Publication No. (NIH) 78–1139, March 1978, 910 pp., and Supplement, *National Institutes of Health Environmental Impact Statement of NIH Guidelines for Research Involving Recombinant DNA Molecules*, Part One, DHEW Publication No. (NIH) 1489, 147 pp., and Part Two, Appendices, DHEW Publication No. (NIH) 1490, 438 pp., October 1977. Volume 3, November 1977–September 1978, DHEW Publication No. (NIH) 78–1843, September 1978, 936 pp., and Appendices, DHEW Publication No. (NIH) 78–1844, September 1978, 608 pp. Volume 4, August–December 1978, DHEW Publication No. (NIH) 79–1875, December 1978, 506 pp., and Appendices, DHEW Publication No. (NIH) 79–1876, December 1978, 456 pp. Volume 5, January 1979–January 1980, NIH Publication No. 80–2130, March 1980, 654 pp. Volume 6, January–December 1980, NIH Publication No. 81–2386, April 1981, 570 pp. Volumes 1–5 (5257 pages in all) are for sale from the Superintendent of Documents, U.S. Government Printing Office, Washington, D.C. 20402, and are available in approximately 600 public libraries of the GPO depository system. (GPO Stock No. for Vol. 1, 017-040-00398-6; Vol. 2, 017-040-0422-2, and two-part Supplement, 017-040-001413-3; Vol. 3, 017-040-00429-0, and Appendices, 017-040-00430-3; Vol. 4, 017-040-00443-5, and Appendices, 017-040-00442-7; Vol. 5, 017-040-00470-2.) Volume 6 is not available for sale.

The United States Congress became interested in genetic manipulation early on, first from the point of view of public safety, but also as a promising new technology that was certain to change the course of mankind. The science policy questions raised immediately became the concern of the House Committee on Science and Technology, and particularly its Subcommittee on Science, Research, and Technology, which was chaired by then-Congressman Ray Thornton.

A principal focus of those policy discussions was the NIH guidelines, which became the authoritative standards for the conduct of gene-splicing research in the United States, and indeed much of the world. As such, they became the focus of the debate over the safety of recombinant DNA techniques, and the Recombinant DNA Advisory Committee (the RAC) became the central debating forum. After his defeat in his bid to become senator from Arkansas in 1978, Mr. Thornton returned to private life as president of Arkansas State University. Donald Fredrickson seized the opportunity to ask him to become chairman of the RAC in 1979.

Commentary

RAY THORNTON

My first introduction to Donald Fredrickson came at a meeting I had with "Tiger" Teague, then chairman of the House Science and Technology Committee, at his bedside in Bethesda Naval Medical Center. Realizing that he was not going to be able to be active on the committee for several months, Teague asked me to chair the upcoming hearings on recombinant DNA. He arranged the meeting with Fredrickson because he recognized how helpful Fredrickson would be as the committee tried to establish its proper role in deliberating this difficult policy area.

Tiger Teague was right. Don Fredrickson was instrumental in developing the concept of the NIH guidelines and for this he deserves great credit. I believe that there were three underlying purposes of the NIH guidelines: (1) to establish a means of communication and review more complete and rapid than normal, (2) to assure that the guidelines were conservative while at the same time allowing research to proceed, and (3) to encourage public participation in decision making.

To a large degree, these purposes were accomplished because of the innovative appratus that was put into place. What the guidelines gave us was a mechanism in which to consider difficult questions of science and public policy with well-established, open procedures; moreover, the guidelines set an important precedent for getting input from the different segments of society. But the real genius of the guidelines was that they allowed progress to be made in an area in which a confrontation had developed between science and public policy, without causing either a significant interference in the progress of the science or a disrespect for the public interest in the fiduciary relationship which attaches to all beneficiaries of federal funding.

The guidelines were not formal regulations. Had they been regulations, or worse yet, statutes, extensive hearings and bureaucratic—if not legislative—labyrinths would probably have had to be negotiated before any changes in them could be made. It was a narrow path which was successfully negotiated.

In Chapter 2, Don Fredrickson suggests that a third generation RAC be formed, an idea that I agree with. It has seemed to me that one of the greatest advances to come out of the rDNA controversy has been the development of a process that eliminates some of the problems inherent in the standard processes of regulation and control. If the mechanism developed to address public policy issues related to recombinant DNA could be applied to other areas of science policy, it would be very beneficial. I don't think we should fail to recognize the value in which may have been a serendipitous discovery.

In thinking about the future, I would like to recall the words of Dr. Maxine Singer who testified before my congressional committee in 1977. In her address, Dr. Singer stated,

> For the future, scientists need to continue, together with federal and local governments, to evolve policies that offer protection from potential hazards and preserve opportunities for discovery and development of safe and desirable applications. Scientists must share their insights into the nature of living things with increasing numbers of people, so the debate can be on understanding rather than fear.

My own experience in Congress has convinced me that there is a need for greater communication between policy makers and scientists. It has become clear that without a common understanding of the issues under consideration, public decisions affecting science will be based on incomplete, or even simplistic information. Under these conditions, the potential for mistakes is high.

The public is still concerned about recombinant organisms, and these areas require continued public input. Recently I received a letter from Congressman Don Fuqua, chairman of the Committee on Science and Technology, and Subcommittee Chairmen Douglas Walren and Al Gore,[1] who express their belief that some scientific experiments should require public review.

Editors Footnote

[1]Albert Gore was elected to the Senate in 1984, leaving his chairmanship of the Subcommittee on Oversight and Investigations of the House Committee on Science and Technology.

> The scientific community cannot isolate itself from the larger community that it serves, and presume that an individual scientist is always capable of deciding that the degree of hazard associated with a particular experiment is acceptable to the public. Moreover, where public funds are spent, the expenditure of those funds carries with it an accountability to the public. Just as the public has a right to regulate the conditions under which human subjects are used in research, so does the public have the right to insure that drug-resistant or toxigenic organisms are not constructed. Certainly, if such organisms are constructed, the public must be assured through some form of open review that they are not released into the environment.

The congressmen believe that some form of research guidelines needs to be kept in place as the technology is further commercialized. They suggest that it may be "prudent to ask the individual IBCs to review any biological hazardous experiment, whether or not it involves recombinant DNA." My own feeling is that further relaxation of the guidelines will occur. But I think we will continue to require a prior review before setting the conditions under which experiments that are possibly dangerous may be conducted, such as developing antibiotic resistance in a known pathogen, or deliberate construction of a toxin-producing organism.

More fundamentally, we should not assume that the communication channels available for resolution of issues involving both science and public policy were so good before the advent of concerns about recombinant DNA, that our effort should now be to return to the "good old days" when scientists were not supposed to be concerned with public policy decisions and public policy could be made to improve the means of sharing information between scientists and policy makers. The need for such sharing of information led to the establishment of the NIH guidelines. Perhaps the lessons learned in this effort might be applied to developed better communication between scientists and the public in other—as yet unforeseen—areas of concern.

The encounter of DNA with Washington politics was a curious phenomenon. It began with some hearings called by Senator Edward Kennedy, first in 1975 after the Asilomar conference and again in 1976 following the formal approval of the NIH guidelines, but perhaps inspired more by Liebe Cavalieri's article in *The New York Times Magazine*. Concerns over the risks of a novel research technique, the prospects of commercial gene splicing, and the limited jurisdiction of the NIH guidelines soon inspired a variety of bills in Congress, most of them dealing with ways to extend the NIH guidelines to all recombinant DNA activities.

The prospect of legislation sent a shock wave through the scientific community, at that time completely unfamiliar with the structure and purpose of such legislation. It was seen as the long arm of "The Law" reaching into the research laboratory and watching every move made by the scientist conducting research. Cries of Lysenkoism and suppression of the freedom of inquiry were heard. But Chicken Little, this time in a white lab coat, was wrong again. Congress, in its muddling wisdom, after two years of contentious debate, came up with a rather benign two-year extension of the NIH guidelines to private research (federally funded research was already covered by the existing NIH guidelines), which never made it to the floor of either chamber for a vote.

Burke Zimmerman, more by chance than choice, found himself in the middle of the maelstrom. In 1976, he testified at Senator Kennedy's hearing for the Environmental Defense Fund. In 1977, at the beginning of the Ninety-fifth Congress, he joined the staff of the House Subcommittee on Health and the Environment, as the subcommittee's science advisor. In that position, he was called upon to be the architect of the legislative proposals on gene splicing for the House of Representatives, subject of course to the fact that it is the elected representatives of the people who ultimately determine policy. But it is the staff that does the grinding, day-to-day work, writes the bills and the committee reports, and is confronted by armies of lobbyists. Here is the account from the inside of what really happened when DNA came to Washington.

3

Science and Politics: DNA Comes to Washington

BURKE K. ZIMMERMAN

Washington is well known as the city of acronyms. For those skilled in deciphering the multitude of cryptograms filling the Potomac air, DNA for many years meant only one thing—the Defense Nuclear Agency. However, around the mid-1970s, particularly after a few popular magazine articles, DNA began to acquire a new meaning in Washington—deoxyribonucleic acid. To many, that sounded even more ominous than its military counterpart.

This was particularly true for whose who learned about DNA for the first time through an article that appeared in the *New York Times Magaine* in August 1976 called ''New Strains of Life—or Death,'' by Liebe Cavalieri, a scientist with impressive credentials. That DNA was the primary chemical constituent of all genetic material was almost an aside to the main purpose of the article. The topic was the new creation of research scientists called ''recombinant DNA,'' whereby the genetic material from different, unrelated species could be combined—or ''recombined''—and expressed in a single organism. The article described horror scenarios in which the recombinant DNA, either by accident or by the hand of some evil genius, would somehow alter the properties of well-known harmless bacteria to produce global pandemics. If exploited by germ warfare specialists, it would produce the most nefarious weapons of all time. At the very least, scientists, in their zeal to exploit a new plaything, would inadvertently disrupt the ecology so as to wreak global disaster, or cause cancer to be transmitted by an infectious agent as is the common cold. There was little mention of the plus side—the doors the new technique opened for understanding the basic mechanisms of life, or the possibility of more immediate benefits, including the cheap production of rare biologicals used in medicine, such as hormones, antigens for the production of safe vaccines, or proteins that regulate the immune response.

The threat seemed horrifying, especially to readers who knew mothing about the science. Oh, some people had heard of Watson and Crick and the double helix. Most had heard of genetics, and they might even have learned somewhere of the concept of a gene. If they had heard of DNA, it was in some benign, esoteric scientific context but certainly not in a way to cause them to give it particular notice. In any case, some of the individuals who first became aware of the existence of DNA, and recombinant DNA in particular, from this article in the *New York Times Magazine* (a most prestigious and respected publication to be sure) happened to be members of the United States Congress. In view of the somewhat frightening way that the subject was introduced to substantial numbers of individuals, what followed during the next two to three years was not surprising. Nor, given the nature of our political institutions, is it surprising that the subject is now nearly forgotten in the halls of Congress.

It is important to note here that the development of safety rules for the conduct of research using recombinant DNA methods had been going on within the NIH for a year and a half prior to the formal release of the NIH guidelines on June 23, 1976. This exercise had resulted from a series of events, including the international Asilomar conference, which followed the 21 September 1973 letter to *Science* from Maxine Singer and Dieter Söll expressing the consensus of the Gordon Conference on Nucleic Acids. It was the scientists, the molecular biologists at the frontiers of research, who first called attention to the many unknowns about the consequneces of putting strange genes into an organism where they didn't normally belong. That a group of scientists had chosen to raise the flag of caution was, in fact, something quite unprecedented. There were many instances in the past when new discoveries were made which posed potential and perhaps even real hazards, yet concerns were not voiced, and little or no self-restraint was exercised. It was usually not until human injury was observed that any measures were taken to control the development or use of a particular kind of technology. And here, the concern was not over the risks of a new technology, but those of the conduct of basic research.

Most of this process had taken place among a relatively small circle of scientists, within the larger community of molecular biologists. Few experts in microbial ecology or infectious disease were consulted initially. There was even a large segment of molecular biologists, including myself, who scarcely knew what was going on. The occasional reports in *Science* about the proceedings of the NIH Recombinant DNA Molecule Program Advisory Committee were interesting but hardly seemed to foreshadow any major controversy. Thus, when NIH called a public meeting of the Director's Advisory Committee to discuss the forthcoming NIH guidelines in February 1976, it took nearly everyone by surprise. Selected individuals were invited to augment the Director's Advisory Committee and present their views. Public interest groups were also invited, but were caught unprepared. In fact, the Environmental Defense Fund (EDF) was the only group that presented testimony at that meeting.

My personal involvement in the ensuing fracas over recombinant DNA began at this time, almost by accident. Because of my interest in environmental problems, I had decided to augment my activities as a research scientist by becoming a part-time staff scientist for EDF. Before I even joined the organization I was told on a Friday night that there was to be a meeting at NIH on Monday having something to do with DNA or genetic engineering, and perhaps EDF should attend and at least meet its responsibility to express an opinion. So that weekend, I read a summary of the proposed NIH guidelines and wrote a two-and-a-half page statement of my reactions, which I typed myself and brought to the open meeting.

It was here that some of the ironies of the whole situation first became apparent. On the one hand, it was an unprecedented exercise; the scientists were really to be commended for their nobility and self-restraint. On the other hand, the federal statutes authorizing any government agency to act were quite limited. Thus, even though the NIH, as requested by a committee of the National Academy of Sciences headed by Paul Berg, had agreed to appoint a committee to draft a set of safety guidelines, they could not impose them beyond their own scientists and grantees. Whether or not NIH should have made what was originally intended by the Berg committee to be a code of good practice into what were essentially regulations for their grantees is another matter. Personally, I don't think so. NIH is not a regulatory agency and should, therefore, have recommended that the entire question of imposing the NIH guidelines as binding requirements be considered in a different, more appropriate forum. In taking the action it did, however, NIH inadvertently enlarged the scope of the debate and intensified it.

First, that the NIH would make observance of the guidelines mandatory for those scientists within its jurisdiction gave the impression that it considered the potential hazards serious enough not to leave safety to the judgment of individual scientists. Second, imposing guidelines only within the NIH immediately created a double standard that, in itself, generated a whole set of issues. Other federal agencies were not subject to the collective wisdom of an NIH advisory committee, nor to the edicts of the NIH director, although most agreed to let NIH take the lead on this matter (see Chapter 2). And certainly, no activities in the private sector need be affected by the internal policies of a federal agency. Therefore, testimony was presented at that first official public meeting from critics who argued that the university scientists working under federal grants perhaps didn't need that much regulation. But if the NIH thinks they do, what about the private sector, especially the drug companies? Shouldn't a legal means be found to apply the NIH guidelines to everyone? It would be the commercial interests that would exploit recombinant DNA technology for everything it was worth. This prophesy is now coming true, on a scale that has surprised even the most imaginative. The question of unscrupulous individuals or nations using the technique for biological warfare was also raised. And it struck many that safety measures that dealt almost exclusively with the possibility of rendering a harmless bacterium infectious to humans while ignoring many other categories of possible hazards were extremely narrowly construed.

So, while many congratulate the scientists and the NIH for their efforts and said that the guidelines were a good beginning, there were criticisms both of the process, which had excluded both the public and a large segment of the scientific community, and of the legal limitations that prevented the mandatory application of any safety procedures beyond the research being supported by the NIH. Thus began a long series of events within the federal agencies and within Congress.

Even with these legal and other problems, it probably would have been relatively simple to resolve safety questions if the body of scientific knowledge had been greater. However, the fact that so much was unknown lent an aura of mystery to the whole process. Even so, if the scientists had been able to agree, the process would have been vastly simpler. But, at this February meeting it was clear that the scientific community was not speaking with a single mind. While most scientists working in the field of molecular biology and gene splicing generally did not want ot see any constraints to their research, they were willing to abide by reasonable safety precautions, even though a majority did not regard the research in question as a hazardous enterprise. However, a small number of other scientists seemed genuinely frightened. Some proposed prohibiting all experiments using recombinant DNA, at least until the risks were known. Others wanted research permitted only under high levels of containment such as that developed at the Fort Detrick research facility when it was developing agents for biological warfare. A leading molecular biologist, Robert Sinsheimer, raised the concern that crossing "natural" genetic barriers between prokaryotes and eukaryotes created a totally unpredictable situation, where even the most educated guesses of the consequences would be of little value. Still others thought that any risk was so remote that the NIH guidelines were unnecessarily, if not absurdly, strict. The arguments on all sides became more intense, as "educated" assessments of probable risk collided, and, because the facts were sparse, could not be resolved. Many well-known and established scientists jumped into the fray, and the polarization of the scientific community, and, hence other groups in our society, was well underway.

By the time Congress began to be aware of what was going on, around the summer and fall of 1976, the scientific community upon which it depended for advice was already somewhat sensitized and divided on the issue. Senator Edward Kennedy had held a hearing in 1975 on the heels of Asilomar. Now he became the first member of the U.S. Congress to respond to the Cavalieri article in the *New York Times*. A one-day hearing was held in September 1976, in which representatives of the scientific community, the federal research agencies, and one public interest group, again the Environmental Defense Fund, gave testimony. As EDF's part-time staff scientist, that duty again fell to me.

The probability of risk was discussed, again with a wide divergence of opinion. The guidelines were defended and the double standard between the public and private sector addressed, with most agreeing that everyone should observe them, but generally avoiding the obvious legal remedy—legislation. "Expert" testimony could be found supporting any viewpoint. Some scientists argued that the benefits to society from such research would be great. Eminent authorities testified that the methods, however carelessly carried out, could not possibly result in any hazard whatsoever; other eminent authorities argued that the slightest slip-up or lapse in precautions could have us all dead.

The Cavalieri article, quoted by Kennedy at the hearing and included in the *Congressional Record* by Senator Jacob Javits, only intensified the latent suspicions that some individuals already had about science in general. The brilliant and mysterious practitioners of science have always been seen as engaging in laboratory sorcery in quest of miracles. But could scientists be trusted? In their zeal to find wondrous new things or perhaps to win a Nobel Prize, could scientists possibly be concerned about something as mundane as public safety? The perception that these scientists, dripping with hubris, would take millions of federal dollars to conduct their research, but would not be altogether honest about what they were doing with that money, inspired a few zealous individuals to set out on a crusade to save the world from the perils of science and the unwitting evil of scientists. The level of emotion reached in the ensuing confrontation bordered at times on uncontrollable rage. And the arrogance of a few scientists in putting down their critics only intensified the mistrust of science held by the worried. Senator Kennedy's hearing did little to soothe the minds of our politicians.

Senator Kennedy did not follow this hearing with the proposed legislation that many anticipated. So two other members of Congress, Senator Dale Bumpers and Representative Richard Ottinger, decided to take action on their own. Inspired in part by the fear and mistrust of scientists that this issue had already generated, the legislation offered by Senator Bumpers and Representative Ottinger was designed as a remedy, requiring all those conducting reserch using these potentially hazardous methods to be mindful of the public safety. They introduced a bill based on the assumption that the conduct of research using recombinant DNA techniques posed a substantial threat to the health and safety of the American people, if only through possible accident. It proposed that all such research should come under strict safety rules presumably to be based upon the NIH guidelines, and that anyone conducting such research would be strictly liable for any injuries that resulted. Violators would be treated much as violators of any of our environmental regulations, being subject to fines and jail sentences. While the legal extension of the coverage of the guidelines to correct the double standard hardly seemed objectionable, the approach was not well thought out and would have created many administrative problems.

In early March 1977, a public forum was held by the National Academy of Sciences. It drew a diverse assortment of scientific antagonists, protests from certain special interest groups, and the news media. Thus it now became time for the major House and Senate committees with jurisdiction over biomedical research to get into the act. By this time, I was working for the House Subcommittee on Health and the Environment, providing counsel on scientific matters to its chairman, Paul Rogers. In view of the media attention the issue was receiving, the Academy Forum and the bill already introduced by his colleagues, one of the first assignments Mr. Rogers gave me was to compose an intelligent legislative approach to recombinant DNA research and to organize three days of hearings on the safety and ethical issues surrounding the proliferation of the new gene-splicing methods.

For me as a scientist, the thought of restricting research unnecessarily with burdensome regulations was offensive. I still had some problems with NIH guidelines, but they were generally reasonable and did not prohibit the kinds of experiments that people were likely to want to do. Even the P-3 level of physical containment, considered by some to be rather onerous, was little more burdensome than the type of precautions scientists doing mammalian cell culture were used to observed routinely. In the case of cell culture, it is necessary to take precautions with one's experiments in order to prevent microbial contamination from the environment. The procedures outlined in the guidelines, however, were intended to keep any recombinant DNA material from reaching the environment and thus possibly infecting a human being. The methodology, however, was similar. Thus, as a starting point, the NIH guidelines provided a sound technical basis for any safety standards. This and all subsequent legislation cited the guidelines and never presumed that the scientific wisdom of Congress exceeded that of the NIH.

The primary motive behind such legislation then, at least in the eyes of Paul Rogers' science advisor, was to correct the legal inequity that permitted safety standards to be imposed only upon NIH grantees. So I got together with the subcommittee's legal counsel (Steve Lawton) and an expert legislative draftsman (David Meade) and we set to work. Citing the rationale of the commerce clause of the Constitution, the NIH safety rules were to be required by all who engaged in recombinant DNA activities within the United States. The logic was simple. If the use of any technique posed a sufficiently large perceived hazard so that NIH deemed it necessary to require certain safety precautions to be observed by someone with an NIH grant, then it followed that everyone else should do the same. However, it turned out that to impose a universal safety requirement was not so simple.

Any such law would have to specify a mechanism for implementation, a seat of responsibility for administering such standards, procedures for promulgating regulatiolns, a mechanism to change them as new data became available, the treatment of proprietary data, the relationship of federal law to state and local authorities, and a means of enforcement. The necessity for the last provision was underscored by some well-publicized cases in the drug industry that indicated that such profit-making organizations could not always be trusted to obey FDA rules to ensure drug safety, even when faced with a federal requirement which imposed penalities upon violators. Any law that failed to address all of these issues would be considered incompetently drafted and would leave the designated administrative agency in a compromised legal situation. Just as the failure of lay critics to understand complex scientific arguments was a major frustration and a principal source of the controversy, so the failure of the scientific community to comprehend the elements of administrative law was equally frustrating and a source of much unnecessary contention. That is, just as many concerned laymen feared the jargon-laden science that they did not understand, so did the scientists fear the law, couched in an ominous jargon of its own. Both of these fears were extreme overreactions to that which was not understood and was therefore, perceived as threatening.

In the meantime, a federal interagency committee, chaired by Dr. Fredrickson, then director of NIH, attempted a preemptive strike. Believing that legislation was a certainty and fearing the worst from Senator Kennedy, the administration drafted its own bill. There was the conviction too that federal legislation was the only way to prevent a proliferation of strict local ordinances, such as that in Cambridge, Massachusetts. Accordingly, a universal observance of the guidelines and proposing federal licenses for research facilities was drafted and sent to the White House for approval. But there the ploy backfired. A former public interest lawyer on the Domestic Council staff insisted that state and local governments should have the right to enact stricter requirements if the citizens so chose. Thus the bill was sent to Congress, and Fredrickson was now in the position of having to endorse an administration position he didn't believe in. Both Rogers and Kennedy introduced the bill as a courtesy. But it was not particularly well drafted and was essentially ignored by the Congress.

The long-awaited legislation from Senator Kennedy finally emerged. It was based upon somewhat different premises, perhaps justifying the fear of legislation. Responding to the horror scenarios described in detail by certain scientists, and perhaps also to a latent mistrust of the scientific community, S. 1217 would have established a national regulatory commission to oversee all recombinant DNA activities in the United States. The proposed legislation, in fact, was patterned after the Atomic Energy Commission Act of

1946, with "Atomic Energy" crossed out and "Recombinant DNA" inserted in its place. Otherwise, the language was preserved nearly intact. This immediately carried with it the message that Senator Kennedy perceived that the potential hazards of recombinant DNA were comparable to those from nuclear fission. In fact, the development of gene-splicing techniques was often compared with the discovery of the means to split the atom, an analogy that produced grimaces of frustration in scientists.

Predictably, as news of the introduction of these bills spread, a furor was created within the scientific community. The legislation introduced by Senator Kennedy that would require the establishment of an entire new bureaucracy to administer research in an exciting new area of molecular biology struck the scientific community as a terrible precedent that would lead ultimately to federal control of the content of science, taking away the scientist's precious freedom of inquiry.

Most scientists didn't really object to the imposition of the guidelines, per se. But somehow, being required by law to observe them was an intrusion of Big Brother into the sanctity of the laboratory. They generally didn't understand the provisions for enforcement. Companies are used to inspection, manufacturing processes long having been subject to one sort of government regulation or another. University-based research is not. Therefore, the boiler plate language lifted verbatim from the Toxic Substances Control Act that was put into the proposed legislation by the House legislative counsel to permit the inspection of research laboratories caused a reaction of some violence. To be sure, the choice of words was not the best. What may have been appropriate for a contentious, somewhat hostile corporation trying to cover up violations of antipollution regulations only insulted the research scientist. And monetary penalities for violators and even criminal penalities in some versions of the bills were met with even more intense shrieks of horror.

For someone knowledgeable of the way federal regulatory legislation is implemented, the provisions of the Rogers bill are more or less commonplace. No large crew of federal inspectors was anticipated. And rarely, even in the most egregious breaches of federal pollution controls, are large fines imposed. The committee report intended to accompany the legislation made it very clear that the penalities were aimed at commercial establishments, not at university research laboratories. Moreover, a scientist working under a federal grant was already required to observe the guidelines. NIH was empowered to cut off the grant of any violator, a punishment that could end one's scientific career. For such a person, passage of the law would scarcely have been noticed. Ironically, the companies hardly flinched. Yet the cloistered scientist, naive in the strange machinations of federal administrative law, predicted the end of the world of open scientific research, soon to be destroyed by mindless bureaucratic oppression.

Thus, the battle lines were soon drawn; scientists began to organize and a few ring leaders emerged, carrying their cause to national scientific meetings. Those science groups experienced with the law and the ways of Washington behaved quite rationally, but there was a general perception in the ranks within the research laboratories that all legislation was the same and that it was all terrible. Few bothered to distinguish between Senate and House bills or to understand the legal nature of the regulatory process. Legislation *per se* was the enemy, to be fought at all costs. It was "the only good Indian is a dead Indian" policy all over again.

But for all the perceptions—or misperceptions—of the majority of scientists, the imposition by law of such safety requirements turned out to be one of the minor issues in the gene-splicing war as it was fought in the halls of Congress. Few disagreed that the NIH guidelines were more or less appropriate standards and that everyone should observe them. One question that came up immediately, but was not expected to be a source of trouble, was whether a local jurisdiction—city or state—should have the prerogative of imposing stricter standards over recombinant DNA research. That seemingly trivial provision of the legislation became the most hotly contested issue for the next two years.

Senator Kennedy saw his role as the defender of the people against powerful special interests, which would otherwise ignore their legitimate concerns over safety. But he was from a state that had such powerful vested interests as Harvard and MIT, which were concerned about a local ordinance imposed by the Cambridge City Council. Its provisions went beyond the guidelines. Harvard spent a fortune lobbying for federal legislation, not to regulate research but to override the Cambridge edict. Yet Senator Kennedy was steadfast in his defense of the right of states and communities to impose their own rules. Paul Rogers' constituents from West Palm Beach, Florida, scarcely knew that there was an issue over gene manipulation. But he saw it as his duty, as the chairman of the primary subcommittee in the House having jurisdiction over biomedical research, to see that things were done right. Always the peacemaker, the prototype consensus politician, Rogers understood the importance of a promising branch of science and opposed needless restrictions. Yet he perceived the guidelines as reasonable and believed it important to create the legal authority to extend them to the private sector. Based on the proposition that a uniform federal standard was appropriate, H.R. 7897, the prevailing legislative proposal in the House of Representatives in 1977, would have required a local government to provide a convincing argument that stricter standards were "necessary to protect health and the environment" before they would be permitted to stand. This solution would have created a procedure whereby any state or community could challenge the adequacy of the guidelines, but arbitrary action that would impede the conduct of science would be prevented. Both Harvard and the scientific professional societies accepted this solution to the problem.

It should be pointed out here that for all its intensity, the recombinant DNA controversy in mid-1977, the time when it peaked, was still a minor issue in Congress. The majority of members of both houses of Congress were unconcerned about the machinations of genes and of scientists, and few had ever bothered to learn what DNA was. Much of Washington, especially in that mammoth polygon across the Potomac, still believed that DNA stood only for the Defense Nuclear Agency. In order for an issue to rank with the top political issues of the day, it must have immediate consequences, it must affect some large or at least powerful segment of the population, and it generally must deal with one of the following: taxes, employment, war, inflation, gasoline, or some other issue touching either the pocketbooks or the hearts (and in a contest between the two, the pocketbooks) of a significant portion of a lawmaker's constituents.

But, among those who were concerned, the battle raged and emotions ran high. Federal preemption was the issue—to be or not to be. Who would have predicted that the battleground for a states' rights issue would be recombinant DNA research? Meanwhile, the scientific arguments, upon which the entire rationale for the guidelines and all else ultimately rested, continued, resolving little. A conference held in Falmouth, Massachusetts, in June 1977 did, however, conclude that the bacterium *E. coli* was probably a rather safe host is which to clone foreign DNA, even though the bacterium's natural habitat is the the colon of warm-blood mammals, including man.

How could congressmen make sense of the brawl that was taking place among scientists? How could one tell whether George Wald and Erwin Chargaff, both aging but distinguished scientists, were right in saying the research should be greatly curtailed because it was so hazardous? Some bright and articulate young scientists such as Jonathan King and Richard Goldstein were saying the same thing. How could a layman analyze the arguments of other scientists such as Charles Thomas and Bernard Davis, who doubted that anything could possibly go wrong?

However, by the way the legislation was shaping up toward the end of the summer in 1977, it really didn't matter who was right. Most reasonable members of Congress who paid any attention to the issue were of the opinion that Congress could not make the technical choice as to whether the use of recombinant DNA methods posed a serious hazard or not. They recognized that a group of scientists had been deliberating for two years to come up with a set of safety guidelines which most scientists considered reasonable. Even those scientists who contended that there was scarcely any potential hazard from the use of recombinant DNA methods generally endorsed the NIH guidelines; some, like Charles Thomas, Jane Setlow and Waclaw Szybalski, were members of the original committee that wrote them. Moreover, even those who considered the technique to be extremely dangerous admitted that the observance of the guidelines would considerably reduce the probability of an accident. Therefore, it was the role of Congress not to judge a scientific question, which it was not equipped to deal with, but rather to establish the political and regulatory mechanism for insuring that research was conducted safely, according to the standards set by NIH, and to enforce those safety regulations in a way that would have meaning. At the same time, even the members of Congress favoring strict controls acknowledged that recombinant DNA methods offered great promise in understanding the mysteries of life as well as providing direct benefits to medicine, agriculture, energy, and the environment. It was further recognized that a mechanism should be kept in place to respond to the changing state of knowledge. That is, regulations could not be written which would last for all time; they must be made flexible to respond to the current state of knowledge.

In view of the above, one would have expected that rational legislation would have had rather clear sailing through Congress. But the outrageous provisions of the Kennedy bill produced outrage from most of the scientific community. Scientists with no experience in the ways of Washington came to lobby, focusing most of their energy on S. 1217. It worked. By the fall most of what little support that ever existed for Senator Kennedy's bill had eroded away. The well-organized, middle-of-the-road scientific lobbies, such as the American Society for Microbiology, generally supported the approach of the House legislation. It was clear that a large majority of members of the House Commerce Committee also would vote favorably for such a bill. But a segment of the scientific community, sensitized by Kennedy's insensitive approach to science, opposed all legislation, per se. Most of these militants never really understood why a bill had to be written in certain ways. So they made little distinction between genuinely bad legislation and that which was competent and, in fact, designed to protect the scientific community from harsh arbitrary controls imposed at lower levels of government.

Then tactics began to get a little dirty. Stanley Cohen of Stanford Medical School, with one of his postdoctoral fellows, did an experiment showing that under certain forced conditions, the recombination of extraneous genes could occur within *E. coli* protoplasts. He then sent a letter to the director of NIH, with copies to leading members of Congress, stating that he had evidence that interspecies DNA recombination was a naturally occurring phenomenon and, therefore, no regulation was needed. Cohen's letter gave Kennedy the excuse he needed to withdraw his bill and save face. Cohen also managed to convince Harley Staggers, chairman of the House Commerce Committee, that even the legislation pending in the House was oppressive and that it should be stopped. Even though the scientific societies endorsed the proposed legislation and the committee votes were there, Mr. Staggers, with letters in hand from Stanley Cohen and Norton Zinder, used his chairman's prerogative to delay the markup of H.R. 7897 on the morning it was scheduled so that it was not completed by the time the clock reached 12:00 noon, allowing Staggers to adjourn the session.

By the time Staggers was persuaded to reconvene the committee, other forces came into play. One of the provisions of the bill that survived to this point was the one on preemption, permitting local standards to override the federal regulations only if it could be shown that stricter standards were necessary to protect health and the environment. This provision became the rallying cry of the public interest groups, who said that this was an impossible burden of proof and lobbied intensively to restore local determination, even it it meant stopping legislation altogether. They commended the Cambridge City Council's actions and did not want to see the rights of a community overridden by the federal government. In the House Commerce Committee, there were perhaps six members who were in sympathy with such a view and strongly supported legislation that would not restrict local activity, even though they knew (at least one would presume they knew) that if the bill were to be defeated on these grounds, there would be no safety regulations whatsoever applying to private industry, which, after all, was the primary purpose of legislation. Harvard University's lobbyists opposed any bill that would weaken the language on preemption, even by a small amount.

Thus, when the committee was reconvened in October to continue the markup, some compromise language was offered in order to calm the stormy waters. Congressman Eckhardt, the great peacemaker, said "Well, 'necessary' might be a little bit strong—why don't we use a softer word?—'reasonable'." This was the word that the public interest lobbyists had agreed to accept earlier. To clarify just who should decide, Eckhardt proposed adding the words ". . . if the Secretary determines that . . ." and Rogers agreed. In the real world, how much difference would it make anyway? Perhaps none, but in the world of perceptions, Harvard's lobbyists felt betrayed that Mr. Rogers would not oppose the compromise and did all they could to stop the bill, going directly to Mr. Staggers. Mr. Staggers, who had a son who had graduated from Harvard, was sympathetic and refused to convene the Commerce Committee for the remainder of the year, for to do so would have meant completing the markup of the legislation and its approval by the Committee.

At that point, another phenomenon was occurring that happens often when difficult legislative issues are protracted. Bills may be introduced that start out with a very simple premise and attempt to do something in a straightforward way. But as each member begins to respond to specific constituent interests—a provision here and a provision there—one sees what I call the "Christmas tree" effect. Pretty soon the bill becomes so weighted down with ornaments that it becomes difficult to see if there is really a tree underneath it at all or at least what kind of tree it might be. This was beginning to happen to H.R. 7897. There were provisions for health examinations for all workers, inspection provisions, provisions to make sure that the trade secrets of industry were not revealed, specifications for exactly what the local biohazards committee should do—not to mention, of course, the issue of preemption, which seemed to be one on which no satisfactory compromise could be reached. The more complex a bill is, the more people will find something wrong with it and thus offer more amendments in the next round of the legislative process. In the final abortive markup of H.R. 7897, there were perhaps another fifteen new amendments being offered. Of course, most would not be accepted. But some would be, and language would be changed yet again. By that time, H.R. 7897 was no longer a bill for which I could feel very much creative pride. It was perhaps best to start over, if that could be done.

Over the Christmas lull before the beginning of the next session, it appeared that the only hope for pulling the legislation out of the morass in which it was foundering was to draft a much simpler bill. With Rogers' support and Harley Staggers' cosponsorship, the critics in both the subcommittee and on the outside might be assuaged. So a simpler bill was drafted (H.R. 11192) and offered as a replacement for H.R. 7897. Mr. Staggers was convinced that it would not suppress free scientific inquiry and would get Harvard off the hook of the Cambridge ordinance. The bill specified relatively few things. The Secretary of Health, Education, and Welfare was directed to draft administrative regulations to extend the standards specified in the NIH guidelines to all, as a two-year interim requirement. The industries worried a little bit that their proprietary interests might not be protected as much as they would have liked, but they generally accepted it. The scientific societies saw the new bill as an improvement and supported it, although questions were raised whether such a position really reflected the views of most of the societies' memberships.

In order that Mr. Staggers would cosponsor the bill, the legislation again included the provision that to override the federal standards, local or state rules must be shown to be "necessary" to protect health and the environment. I was perhaps naive to think that the controversy over preemption had cooled off over the winter. Since the issue was not resolved, it was predictable that the antagonists would again come to blows over it. But the votes were there. The Subcommittee on Health and the Environment, followed by the full Commerce Committee, quickly passed the proposed legislation by a large majority. Six Congressmen, all of them Democrats (Waxman, Ottinger, Markey, Maquire, Mikulski, and Gore) voted against the bill and filed a minority report, stating their objection to the bill primarily on the grounds of the federal preemption issue. The Committee on Science and Technology, asking for sequential referral, passed an almost identical bill.

The Senate, or at least Senator Kennedy, had been forced to back down from the fiasco of S. 1217. Larry Horowitz, Kennedy's principal aide on health and science issues, said that the Senate would wait and see what the House did. Kennedy introduced a version of H.R. 11192 which again gave the state and local governments the power to override federal regulations. If the House passed legislation, Horowitz said, Senator Kennedy would sponsor it in the Senate. If not, they weren't going to touch the issue. They had failed taking their own initiative and gotten burned. Senator Kennedy was not about to risk losing face again.

It was mid-1978. The House of Representatives requires that a "rule" be granted to govern the floor debate and the way a bill will be voted upon.

A rule was requested but the Rules Committee never acted. And so, like so many other pieces of legislation before the 95th Congress, H.R. 11192 died an ignominious death in obscurity. That same Congress failed to pass many bread-and-butter issues, including the re-authorization of many uncontroversal programs. So it was not so much a personal blow to my ego that the recombinant DNA bill, on which I had worked so hard, didn't pass either. There are many false perceptions of why the bill failed, particularly those of some scientists who believe that it was their lobbying that killed the legislation. What really did it in was that the attention span of both the defenders and the opponents of a basically sound piece of legislation was all but over by the time Congress adjourned on 15 October 1978. More than two years had passed since Senator Kennedy's 1976 hearing. Industry had not been under any rules even though they were beginning to explore and exploit the new technique. There had been violations of the guidelines reported in university labs, but no one had even caught a cold or gotten a stomach ache.

How long can individuals maintain a zealous fervor over an issue that can't even be shown to be real? Well, that experiment was carried out during the gene-splicing wars. The results showed that the period of intense interest certainly does not exceed two years and may well be somewhat shorter. Ironically, the individuals in public interest groups who lobbied for strict, complex legislation wound up preferring no legislation to any that contained even conditional federal preemption. Even Pam Lippe, then representing Friends of the Earth, who was perhaps the most determined of all the public interest crusaders, I had met, tired of the issue and found that it was much more fun to fight nuclear power than recombinant DNA. So recombinant DNA had its day in Congress. It was a novelty, with ominous implications. But like all novelties, it wore off.

Gene splicing became an issue in Congress when certain of its members became aware, with the help of articles like Cavalieri's, of what they perceived to be a frightening new power in the hands of scientists. Some began to look at the scientific community with a measure of fear and suspicion. It was as though some watchful citizens had brought to their attention a group of curious but selfish children playing with dangerous explosives. It was up to our collective parent—the U.S. Congress—through a few champions, to see that they didn't blow us all to Kingdom Come. But alas, even though the unruly child was never disciplined, the awaited explosion never came. So the principal, or at least initial, motive for congressional concern faded away. The question of safety was, and still is, a valid question. The scientists couldn't agree on precisely how risky splicing might be, but then, they really didn't have to. Whatever the intrinsic hazards in mixing genes may be, they seemed neither immediate nor life-threatening. And who really cares about taking the trouble to create a legal mechanism just to insure consistency between government and private research? Is it worth the time and trouble, especially when there are far more important political issues? Let's face it. Without the *real* threat of plagues and monsters, the recombinant DNA safety issue is downright dull. As proof of this assertion, the recombinant DNA registration bill, introduced by Adlai Stevenson in 1979, could not excite anyone, either for or against it.

Without the ominous and mysterious overtones that surrounded the subject in the beginning, legislation of the Paul Rogers type may have been introduced anyway. Without the notoriety and publicity over the issue, it would then have been a standard, somewhat drab, congressional issue. It is likely that it would have passed easily, and perhaps unnoticed. Without the sensitization that had already aroused the scientific community, would scientists really have cared if the NIH guidelines, which already governed most university based research anyway, were simply extended to private industry, as long as the law were written to permit changes to be made quickly and simply? Such legislation wouldn't have had a noticeable effect on the NIH-supported molecular biologist. Industry is used to being regulated and never objected very strongly to the bills that would impose the guidelines upon them anyway.

To reflect now on the anxiety that gripped scientists, civil servants, and a small but very agitated segment of the public reveals certain truisms about human nature that we have no doubt seen in many other contexts. First, the *belief* in a set of premises is a much more compelling force than a probable or even a proven set of facts. Second, without something tangible upon which to focus one's attention, no one can maintain an intense interest in any issue for more than a limited time. Third, politicians can't be bothered with esoteric issues, at least not for long, especially ones that are unimportant to their political futures.

What, then, may we expect to see in the future? It is easy to predict that Congress will continue to be captivated by the marvels of science and the technology that it spawns, particularly if it results from a federal program. But I would very surprised if anything quite like the recombinant DNA controversy were to occur again for some time unless there is a clearly demonstrated hazard that is somehow both greater and uniquely different from common diseases or other involuntary hazards. Even then, it was the *unknown* risks that generated much of the melodramatic aura that surrounded the discussions of genetic manipulation. People are generally much more comfortable with hazards that are known and understood, even if the danger is great (nuclear weapons) than they are over the unknown, even if it may be relatively harmless (genetically engineered bacteria).

I suppose it is possible that there could again be a scientific issue that raises concerns over public safety. But the Wolf Principle (or perhaps you prefer the Chicken Little Principle) may prevail. Several years may be required before we have forgotten enough of the recombinant DNA debate to be willing to go through something similar. Or at least, the issue may require a new set of actors—new scientists, new politicians, and a new set of enthusiastic individuals with a cause to champion. In any case if a similar controversy does occur, it will again probably have nothing to do with the actual degree of risk.

Recently, one individual has been trying to generate a new public issue over genetic engineering. In September 1983, Jeremy Rifkin, a perennial and strident critic of genetic engineering for the past decade, filed suit against the government to prevent field trials for a genetically engineered strain of bacteria lacking the protein that produces ice nucleation on plant surfaces. He contends that there needs to be greater scrutiny of genetically engineered organisms released in the environment. Many agree that it is valid to require an assessment of the ecological safety of organisms to be released into the environment and assurance that such organisms will not pose a threat to the ecology. But he seems to have picked the wrong subject as a test case, for nearly identical organisms have exist in nature and have probably been around for a long time.

Most observers regard this as a nuisance suit and little more. Rifkin's support has been largely from the religious community, and even that has been eroding. The experiments in question have already been done with identical bacteria arising in nature and produced by conventional genetics; the RAC readily approved the field trials for the genetically engineered equivalent. In a 1984 decision, U.S. District Court Judge John Sirica supported Rifkin on technical legal grounds only, in his contention that the experiment fell under the National Environmental Policy Act (NEPA), but refused to hear Rifkin's witnesses ready to present scientific arguments that it posed a danger to the environment. The RAC subsequently approved an identical experiment to be carried out by a private concern, thereby not falling under the jurisdiction of NEPA. In appealing the Sirica decision, NIH agreed to submit an assessment of the possible environmental consequences of each deliberate-release experiment it approved. But the U.S. Court of Appeals overturned the decision that would have required NIH to write a programmatic Environmental Impact Statement (EIS), in the way that the law requires be done for nuclear power plants or hydroelectric dams. But for all of Rifkin's self-generated publicity over the issue, and his appearance on numerous television talk shows, there is simply not enough interest for another national controversy.

Is the Congress likely to get involved again, especially in issues that are not presented in the sensational and melodramatic manner that characterized the first public accounts of gene splicing? So far, there has been a hearing on the safety of releasing genetically engineered organisms into the environment. But Rifkin has not inspired anyone in Congress to introduce new legislation. Will an appropriate federal agency take action on its own? The Environmental Protection Agency (EPA) has announced plans to regulate the release of genetically modified organisms into the environment under the Toxic Substances Control Act, a process likely to require some years for hearings and the writing and evaluation of proposed standards. Such actions will generate much concern among the biotechnology companies pursuing agricultural products. But this is the way most regulatory issues are dealt with; these are not the ingredients of another civil war.

But suppose that another situation arises that produces concerns comparable to those scientists had over the new recombinant DNA technique that they had developed? Will scientists keep quiet for fear that another national spectacle might occur as it did over recombinant DNA? Among the scientists who first called attention to the potential dangers of gene manipulation are many who regretted their actions in view of the public controversy that resulted. It was indeed ironic that in simply exercising their civic responsibility, they brought down upon their heads a deluge of criticism that threatened, or so it seemed, the scientific enterprise itself. Nobel laureate James Watson recanted his once cautious position in a dramatic statement at a public meeting at NIH. It is possible—even likely—that many scientists will be very careful to limit expression of their concerns over any new issue to trusted colleagues.

But even if scientists do make their worries public, what may we expect in view of the way our political institutions have responded to proven hazards? The Reagan administration attempted to dismantle the regulatory apparatus that controls the manufacture of toxic chemicals and their release into the environment, but the public, Congress, and many dedicated civil servants objected. It was too much, too soon. Nevertheless, the 1980s have seen a slow but steady weakening of a number of environmental standards. The regulation of hazardous substances is likely to remain primarily political, rather than be based on technical or medical criteria. Cigarette smoking, for example, the single most serious cause of ill health in the civilized world, seems to be in little danger of being regulated, at least on the federal level. The powerful tobacco lobby has friends in Congress, and presumably the decision to risk lung cancer is a matter of free choice. The enormous health burden imposed by smoking costs everyone, including taxpayers, billions every year. But that, for some reason, is more acceptable than regulation that might impose an economic burden on a single industry. If Congress does choose to look at any issue seriously, the special interest groups will compete and those with the most economic and political power will win. As the biotechnology industry grows and becomes more powerful, will it, too, become another special interest force, able to exert the traditional type of torque on congressional policy that other large industries now do? Moral principle, and that includes the public safety, in spite of what some politicians may say, rarely has much to do with the resolution of the issues before Congress.

Meanwhile, throughout the intense period of controversy and up to the present, the NIH guidelines have been in place and seem to be working.

However, relatively soon after they were instituted, as the use of recombinant techniques became commonplace, the cataloging burden alone—keeping records of all ongoing projects—began to grow enormously for NIH. Some of the provisions of the old proposed House legislation were ultimately incorporated into the guidelines, such as putting much of the administrative responsibility into the hands of local institutional biosafety committees.

With time, as people have become inured to the possibility, however remote, of potential hazards from gene splicing, the guidelines have been continuously relaxed. Years ago, it became an unpublicized goal of NIH Director Donald Fredrickson to gradually phase out the NIH guidelines entirely, at least as mandatory standards, as the public and scientific perceptions of safety permitted. However, as he discusses in Chapter 2, the writing of rational guidelines was part of NIH's central and necessary role in the recombinant DNA affair, and he still sees a continuing and important role for the RAC, or its successor. But all along he maintained his belief that NIH should stay out of the regulation business. And he represented the feelings of many in opposing, at least privately, the bills before Congress that would impose the NIH safety restrictions on industry, not because he believed in a double standard but because he feared that Congress might do something mindless and arbitrary, creating ponderous, inflexible, and inappropriate regulatory machinery that would be ultimately detrimental to science. By keeping the guidelines under the control of NIH, it is possible to change them easily and relatively quickly. Genuine federal regulations, on the other hand, can require as much as eighteen months for amendments to be put into effect. Perhaps a greater advantage in keeping the RAC and the authority over the guidelines within NIH is that the shifting sands of Washington politics have essentially been kept out of the picture. This has made it possible for the RAC's decisions to be based primarily on scientific considerations and not be significantly influenced by political pressure.

The NIH guidelines have served their purpose. Even in the absence of a law, private industry has generally compiled with their provisions and has cooperated both with the procedures of the RAC for reviewing new experiments and with the establishment of biosafety committees. Many now believe there should be only recommended guidelines, and that strict observance should not be a condition for receiving NIH funds. That will probably not happen in the near future. The NIH believes that the RAC should continue to assess new types of experiments for safety and remain the principal forum in the United States government for the discussion of the new issues that continually arise in this rapidly developing scientific area. In 1984, for example, it appointed a subcommittee to consider the ethics of human genetic intervention.

The gene-splicing wars were waged in part by a diverse group of individuals, most of them with scholarly or scientific credentials, whose consciences would not allow them to sit idly by while a revolutionary development in molecular biology was pursued at full speed at the possible expense of the public safety. Some subset of this group of critics, which perhaps numbered forty or fifty people at the most, would turn up the speakers' platform of each of the numerous public forums where the issue was debated. Most approached their mission with a great deal of energy. Many were articulate and persuasive. It was really from this group that the first raids and attacks in the gene-splicing wars were launched. It was the scientists, even those who were writing the guidelines for the RAC, who were put immediately on the defensive. For several, planning and waging the battles became a full-time occupation, and certainly changed the direction of their professional careers.

Where are they now and what are they doing? Have they gone back to what they were doing before this all happened? Are they continuing to evolve their own versions of social criticism or advocating a particular solution to the perils of too much technology too soon? Well most, it seems, belong to a somewhat loose organization called the Committee for Responsible Genetics. The committee publishes *Gene Watch*, a collection of articles and short news items, ranging from strident advocacy pieces and dogmatic exposés of some alleged misuse of modern genetics, to well-written, objective, and highly professional discussions of some difficult subjects. The membership includes two contributors to this volume (Krimsky and Nader), and many others whose names are mentioned frequently throughout these pages. As a group, there has been a definite mellowing, as there has among the other antagonists in the gene-splicing wars as well. The level of emotion of the late 1970s is much lower, and there is a generally more passive and thoughtful approach than that that characterized the days at the peak of the battle. Jonathan King, one of the most active of the critics from the scientific community, has turned his talent for public persuasion to trying to divert the swelling military budget to socially responsible ends. Susan Wright, who still attends meetings of the RAC and who has been working on a book on the subject for several years, has been writing most recently about biological warfare (*Bulletin of the Atomic Scientists*, April 1985). Yes, the people are still there. But the mood has definitely changed. And so has the focus of their concerns.

The fears of a genetically based hazard to human health have been supplanted almost entirely by speculation over a possible threat to the ecology, and even that, as discussed above, runs little risk of assuming the bizarre dimensions of the gene-splicing wars. Rather, the concerns of the "critics"—and the philosophers and the defenders of social justice—are now aimed at the consequences of the applications of the new genetics. Biological warfare waged with agents designed and synthesized by government scientists is only one obvious misuse. Others, such as Marc Lappé (*The Broken Code*, Sierra Club Books, 1985), worry that the priorities being set by the commercial biotechnology industry are not responsive to the needs of the people. That is, the argument is advanced that if the tools of biotechnology are capable of relieving numerous aspects of human suffering, it is incumbent on society to direct its technological resources to these ends, regardless of the "market potential." And many are trying to grapple with the perplexing ethical issues that arise along with the capability of altering the genetic constitution of human beings. The ethical questions arising over the first contemplated attempt to correct human genetic diseases are difficult enough. But these difficulties become compounded when one considers that it is possible to alter the genes of a fertilized mammalian egg so that the changes are passed to subsequent generations. Given the rate of advancement that we have learned to expect in the last fifteen years of molecular and cell biology, the day can be forseen when numerous human traits may be subject to directed manipulation. What then?

Congress began its scrutiny of the ethics of certain types of medical research in the mid 1970s with the creation of the National Commission for the Protection of Human Subjects of Biomedical and Behavioral Research. The commission, the result of a legislative compromise between Senator Edward Kennedy and Congressman Paul Rogers, consisted of a body of interested and educated men and women bringing their diverse perspectives to bear on the problem. By and large, it was a smashing success, in large part a result of the intellect and dedication of its chairman, Harvard physician Kenneth Ryan, and a first-rate staff headed by attorney Michael Yesley. It produced a series of reports on a number of issues, including fetal research, research on children, research on prisoners, and psychosurgery. As the commission's expiration day neared, the Health Subcommittees of both the House and the Senate (again Rogers and Kennedy) considered the type of body, if any, that should succeed the national commission. As Rogers' resident scientist, it was a challenge that I welcomed. It was clear that the ethical implications of many aspects of medical practice and biomedical research were increasing, along with the power of the technologies that were being developed. Thus, even though we were still trying to figure out how to get a sensible bill on recombinant DNA through Congress, and certainly had our hands full, I thought the matter of a new ethics commission deserved careful consideration.

Thus was born the President's Commission for the Study of Ethical Problems in Medicine and Biomedical and Behavioral Research. Again, it was necessary to reach a compromise with Kennedy, but Larry Horowitz and I managed to work out our differences with relative ease. Unlike the recombinant DNA legislation that was stalled forever in the "Twilight Zone" at the House Rules Committee, my efforts were to be rewarded, perhaps because the issue received virtually no publicity or debate. The commission bill passed both houses of Congress on the day of adjournment of the Ninety-fifth Congress as an obscure addition to a much larger legislative package under a suspension of the rules. Congress is much more likely to grant such a waiver and whisk legislation through when it is in a hurry to adjourn and go on vacation. That is the way that some of the best (and worst) legislation gets passed by Congress. Had the President's commission bill been subjected to the full benefit of the legislative process, with committee markups, lobbyists, and enough time for people to realize that there was something to object to, I am sure that it would have suffered the same fate as H.R. 11192.

The commission was given the opportunity to sit back and reflect on the ethical principles that underlie a somewhat greater scope of issues in medicine and research than its predecessor. By the time the legislation was finally implemented, the commission had a lot of work to do in a short time. The conservative Ninety-seventh Congress did not extend its lifetime, except to allow it to finish writing its reports. But during its brief existence, it did take a close look at the future of genetic engineering with respect to human genetic manipulation. In its report, *Splicing Life*, the commission endorsed research that would improve our understanding of the complexities of life processes, including that with potential direct applications to human beings. The commission pointed out that the use of a particular method—for example, gene splicing—did not change in any unique way the ethical considerations that should always apply to research on or affecting human beings. It did recommend that the federal government continue in a formal way ongoing oversight of all proposed applications of genetic engineering involving human beings. But the concept of establishing a conscience as an integral part of the collective governmental brain is unprecedented. Already, proposed legislation creating such an oversight commission has been put off for more than two years. And, like the earlier legislative proposals to subject recombinant DNA activities to federal standards, the effects of political modulation have already altered the concept proposed by the President's commission.

There will undoubtedly be new issues in science and technology—controversial issues that capture some portion of the attention of the public and Congress. The extent of the interest in and, eventually, the resolution of these issues, will depend primarily upon many things—the mood of the times, the prevailing politics and social values, and the effectiveness of the advocates on either side of the question. But not upon the views of learned commissions. When the basis of controversy is a clash between differing ethical values and moral principles that cannot be derived from a generally accepted truth, the resolution of an issue becomes vastly more complex than when the primary focus is on the assessment of risk. The abortion issue is a case in point. As we all know, the prevailing values of any society vary with time, further complicating the task of those who believe that we must begin now to define an ethic to govern the development and use of new technology. Nevertheless, this task is an extremely important one. The power of technology has increased enormously in recent years in its ability to alter drastically the character of the planet and the human societies that inhabit it. We cannot wait until the next equivalent of the recombinant DNA controversy to begin to deal with this matter in earnest.

Sometimes we in America forget that we share the world with a great many other human beings and a good deal of activity in those areas which we somehow come to think of as American. Recombinant DNA was one of those phenomena that, to many, seemed uniquely American, although that, of course, was hardly the case. It is true that the initial successes in the splicing of genes from different organisms did take place in the United States, and that in the early 1970s there was certainly more activity in America than in any other country. But science and the knowledge it generates are universal. And certainly from the time of Asilomar, the concerns over the risks of making genetic chimeras had spread around the world.

Each country had a somewhat different way of handling the matter of safety procedures and the legal mechanism to implement them. In the United States, much of the controversy stemmed from the fact that the NIH guidelines could be enforced only for research sponsored or conducted by that agency. Other federal agencies went along with them, but there was never a mechanism established to extend them to any other entities, in spite of all of the fuss over legislation.

Different models were chosen in other countries, but most looked toward the NIH and America for guidance. In Great Britain, however, it was a bit different. There never was a double standard to become part of the debate, as there was in America. Researchers were beholden to the Genetic Manipulation Advisory Group (GMAG). But there were other concerns. Keith Gibson who served as the Secretary of the GMAG from its inception in 1976 until its dissolution in 1984, and he details these concerns, both in Great Britain and in the European Community.

4

European Aspects of the Recombinant DNA Debate

KEITH GIBSON

In this chapter I want to give a broad overview of the initial reaction to the introduction of recombinant DNA technology in Europe and how the various governments and the European Community responded. However, I will concentrate on the U.K. Genetic Manipulation Advisory Group (GMAG), which evolved its own unique scheme for the assessment of categorization of experiments and continues to remain in many respects independent of other European countries which have chosen to follow the U.S. National Institutes of Health (NIH) guidelines. (I also want to make it clear at the outset that the views in this chapter are expressed by me as an individual and do not represent the views of either GMAG or the U.K. Medical Research Council.)

The Effects of the Initial Expression of Public Concern on Scientific Programs in Europe

It is probably fair to say that when concern was first expressed in 1974 about the conjectural biohazards of recombinant DNA research, United Kingdom scientists were among those most affected in Europe. The scientific programs and the number of U.K. centers involved were more numerous and, therefore, the effect more profound in the United Kingdom than in the rest of Europe.

Because the recombinant DNA technology was restricted in its early development to those centers doing basic academic research in molecular biology and almost exclusively supported by public funds, it was a relatively easy matter for government officers to require those concerned to suspend all work involving the technique until government guidelines for rDNA work had been promulgated. All of the U.K. centers concerned at that time were funded through the Medical Research Council, the Agricultural Research Council, or the University Grants Committee, all organizations that receive their funds from government sources. The government department responsible for providing those funds and to whom the research councils and the University Grants Committee are accountable is the Department of Education and Science (DES). It was this department that therefore immediately responded by, as inevitably any government does, setting up a committee (Ashby Working Party[1]) to advise them whether or not rDNA work should be done in the United Kingdom. Following the Ashby recommendation that rDNA work should go ahead provided adequate safeguards were established, the government set up another working party under the chairmanship of Sir Robert Williams to prepare a code of practice for work involving genetic manipulation. Although both the Ashby Working Party and the 1976 Williams Working Party realized the enormous pressures building up within the scientific community, they reacted with remarkable speed—each reporting within a year of being set up. However, there was considerable agitation in the scientific community because it felt that two years was an inordinately long time to hold up such important work, particularly as the Ashby report had emphasized the—potentially—considerable benefits of the work.

In addition to preparing a code of practice, the Williams Working Party recommended[2] that an advisory body be set up to maintain surveillance of work involving genetic manipulation, and they also advised that statutory regulations be promulgated. The Genetic Manipulation Advisory Group (GMAG) was set up by the Department of Education and Science in 1976 and held its first meeting in January 1977. The GMAG membership has eight scientific and medical experts, together with four members able to represent the public interest; four nominees of the Trades Union Congress, representing the workers; and two representatives of management, one nominated by the Confederation of British Industry—representing the interests of industry—and one by the Committee of University Vice Chancellors and Principals. The decision by the government to appoint members of the public to an expert committee was a unique departure from the usual practice and was regarded to some extent as an experiment to determine whether it should be more widely adopted. The experiment was designed to open up committee work to allow more public participation and a greater flow of information to the public. The members appointed to represent the public were, and have been ever since, experts in a variety of disciplines closely linked with science, such as medicine, philosophy of science, and scientific journalism. Although the contribution of the public representatives has been recognized by the GMAG as important, it remains debatable whether the public members would have been as effective without their individual expertise. The addition of public representatives to expert committees does not appear to have been a widely accepted principle. It may also be of interest to note that GMAG decided that its meetings should be held in private, unlike the NIH Recombinant Advisory Committee (RAC) meetings, which were open to the public. By this time U.K. scientists, who had been advised to stop all work involving this technique until guidelines had been prepared, made strong representations to GMAG and the Medical Research Council because they were aware of some work involving rDNA techniques being carried out in the United States. There appears to have been an unfortunate misunderstanding by officials in the United Kingdom who failed to appreciate that there had been voluntary suspension of work in the United States on certain types of investigation only and not, as they seem to have imagined, a suspension of all work involving the technique. It was even said by some U.K. scientists, who had also apparently misinterpreted the U.S. voluntary suspension agreement, that the voluntary halt in the United States had been engineered by some groups to deliberately gain advantage over other countries.

The GMAG initially accepted the Williams report[2] together with the safety guidelines it contained, and agreed that in order to expedite matters they would follow these closely when reviewing individual proposals to do work involving rDNA. These guidelines placed greater emphasis on well-established physical containment that had been devised for work involving known dangerous pathogens, and good working practices, as compared to the NIH guidelines that placed greater emphasis on biological containment with relatively less stringent physical containment requirements. The Code of Practice in the Williams report recognized four physical containment categories, I (least stringent) to IV (most stringent). Category I equates approximately to the NIH P2 facility requirements and Category IV is more demanding than the NIH requirements for its highest containment category. The categorization of experiments outlined in the Williams report is based on a scheme reflecting evolutionary relatedness, that is, the closer the animal source of the DNA insert was to that of man the higher the containment requirement would be for that particular experiment. This scheme was, as will be described later, abandoned in 1979 in favor of a risk assessment scheme. As the NIH had not promulgated its guidelines at the time the Williams report was issued, there was some satisfaction gained for U.K. scientists in that they could now go ahead with their work—even if at a relatively high containment level and with each experiment reviewed on a case-by-case basis by GMAG—and possibly regain some of the ground they felt they had lost to U.S. scientists.

The Medical Research Council in its turn agreed to provide, exceptionally, funds as a matter of priority for laboratory containment facilities not only in its own units but to university departments. (This was a unique departure from the council's funding policy, as accommodation and basic equipment for laboratories is usually provided throughout the University Grants Committee as part of the dual funding[3] arrangements with council.) The Williams guidelines placed much of the work that was of interest at that time into Category III. The average cost of constructing a Category III containment facility in 1976 was £70,000, in addition to the relatively high cost of running and servicing such a facility. About ten such facilities were constructed in the United Kingdom and paid for by public funds in some way or other.

Although the GMAG's terms of reference were very wide, it decided early that it would never wish to be involved in research policy matters and would not therefore operate any form of peer review system. It was felt that almost all work would be subject to some form of peer review whether it be at the national level (that is, through research councils and other funding bodies) or locally through scientific review committees. The GMAG's principal concern was safety—for the worker and the public.

Safety Consciousness

It may be appropriate to reflect on the general political situation in the United Kingdom regarding health and safety. The growing demand for increased health and safety at work and a greater awareness of the many risks involved in any activity, which was occurring in the early and mid-1970s, heavily influenced the course of events both in the United Kingdom and the rest of Europe on the approach and attitude to rDNA research.

At the start of the rDNA controversy the United Kingdom had a Labour government which gave high priority to health and safety at work, and this together with its close relationship with the trades unions meant that a very comprehensive Health and Safety at Work Act was introduced in 1974. This act gave much broader powers to the government's Health and Safety Commission, through its Health and Safety Executive (HSE) and their Factory Inspectorate. The HSE Inspectorate may visit any place of work to ensure that safe practices are being followed and regulations complied with, although government laboratories and hospitals are not bound by these regulations. However, the government has agreed to comply with all the statutory provisions and to allow inspectors to visit their premises. There followed very quickly special statutory regulations under the Health and Safety at Work Act which required all establishments to set up local safety committees. I emphasize this for three reasons: (1) universities and research institutes were covered by the Health and Safety at Work Act; (2) it strongly influenced the way GMAG operated,; and (3) it influenced the attitude of other countries in the European Community, as will be discussed later.

So not only were research institutes and other work places required by law to set up local safety committees, they were also required by GMAG to set up local biological safety committees to discuss work involving rDNA activities. The local biological safety committees were not—and are not—a statutory requirement, but GMAG will only give its advice when proposals have been discussed with the local biological safety committee. Such close scrutiny of experimental protocols before work had even started, and the intense surveillance of all work, but rDNA work in particular, were unprecedented.

It should also be noted that under the genetic manipulation regulations it is the Health and Safety Executive that inspects premises and keeps them under surveillance (this includes both university and industrial laboratories), although GMAG has always visited all new category III and IV containment facilities before giving advice on such facilities and the work to be done in them. The HSE routinely accompanied GMAG on such visits as observers, but it was made clear that these were not formal HSE inspections. HSE and GMAG have always enjoyed a good working relationship.

Training

It was of concern to the GMAG that some of the centers initially set up to do rDNA work were biochemistry and molecular biology laboratories which had no previous experience in either the handling of microorganisms or the operation of biological containment laboratories. Laboratories that lacked the necessary microbiological experience were encouraged by GMAG to send members to appropriate training courses or collaborate with departments having the relevant experience.

At the GMAG's request the Medical Research Council sponsored training courses for biological safety officers and deputy biological safety officers at the Microbiological Research Establishment (Porton)—now the Public Health Laboratories Service Centre for Applied Microbiology and Research—on the operation of containment facilities. Following the introduction in 1979 of the GMAG assessment scheme, most experiments were recategorized to the lower containment level (that is, Category I) or "good microbiological practice.'' It has therefore been agreed that courses on operating containment laboratories specifically for rDNA work are no longer required.

Scientists' Concerns

Among the initial concerns of U.K. scientists about the introduction of GMAG and its procedures were the following:

> Overzealous safety committees would cause unnecessary delays and interference in their work.
>
> The safety committees would demand unnecessary and expensive safety requirements which would have to be paid for from dwindling financial budgets.
>
> Disclosure of details of each experiment to a large number of people (either locally to biological safety committees or nationally through GMAG and the HSE) at an early planning stage might give competitors an opportunity to steal their ideas or, at the local level, subject all work to an uncontrolled peer review system.

Many of these fears were unfounded, but until the GMAG procedures had been put into operation and the HSE's role clarified, there was considerable suspicion. Much of this dispelled following the introduction of the GMAG risk assessment scheme in 1979. Following the full implementation of the GMAG risk assessment scheme in January 1980, some rDNA work has been exempted from the genetic manipulation regulations and almost all work may now be done at the lowest containment level (Category I) or ''good microbiological practice.'' Categorization of experiments is carried out by the local biological safety committee and only in cases of uncertainty or where work in Categories II, III, or IV is envisaged is GMAG asked to advise on a case-by-case basis. (Up to 1980 each experiment was considered by both GMAG and the local biological safety committee on a case-by-case basis.)

It is interesting to note that in the early days many of GMAG's most vociferous opponents were scientists from academic and industrial laboratories who were not even involved in rDNA research. I find it disturbing that eminent men should be so openly critical without offering alternatives for consideration or openly presenting a scientific case for public information. Such criticism seemed likely to deepen the public mistrust that some people may have had about the work.

The Royal Society of London set up an ad hoc group[4] to discuss the safety and policy aspects of DNA work. The Council of the Royal Society issued a statement prepared by the ad hoc group criticizing GMAG and its risk assessment procedure for the categorization of experiments (*Nature*, 1979).[5] The Royal Society[6] at that time urged GMAG to follow the NIH guidelines and also sponsored (with the Committee on Genetic Experimentation)[7] a conference on rDNA held at Wye, Kent, in 1979 (the proceedings were published by Pergamon Press the same year). Though many were critical of GMAG's approach to categorization and the difficulties that might be caused in terms of international collaboration, the differences between the GMAG and NIH guidelines have always been rather more apparent than real. The difference also reflected nationalistic differences that took account of the various shades of opinions and political backgrounds between the two countries. It should be pointed out that even in those countries that opted to follow the NIH guidelines, there were operational differences and, in certain countries (Holland, for example), constraints placed upon their implementation.

When it was realized by GMAG that much work was being downgraded to "good microbiological practice" (GMP), it was noted in January 1980 that—in the United Kingdom at least—there was no definitive statement on what GMP was. Scientists appreciated that GMAG was having difficulty in reaching a decision on GMP and a Joint Coordinating Committee for the Implementation of Safe Practices in Microbiology formulated a guidance note entitled "Guidelines for Microbiological Safety" and which is reprinted in GMAG's third report.[8] This document was accepted in July 1980 by GMAG as a suitable basis for GMP.

In my view the Joint Coordinating Committee's note on GMP illustrates—almost exceptionally in the rDNA debate in the United Kingdom—a constructive independent initiative taken by the scientific community, collectively, to facilitate in a most helpful way the work of GMAG. On an individual basis, Dr. Sydney Brenner (of the Medical Research Council's Laboratory of Molecular Biology) first gave, in July 1978, the GMAG the initial concept for the risk assessment scheme that proved so successful. It was introduced in March 1979, and revised in January 1980. Local biological safety committees were able to operate with ease the risk assessment scheme, which was sufficiently flexible to allow scientific or medical information to be introduced when reaching a decision. The scheme resulted in most work being recategorized to either Category I or GMP. It was also Dr. Brenner's development of the disabled Medical Research Council (MRC) *E. coli* strains (MRC 8 in particular) which allowed GMAG to incorporate the concept of biological containment in its risk assessment scheme with greater confidence. One should acknowledge here the enormous individual scientific contributions made to the GMAG's various committees and their expertise and guidance which was so vital to GMAG's work.

Impact of the GMAG in Europe

Because the U.K. Williams report and GMAG Code of Practice appeared before the NIH guidelines, those countries in Europe involved in the work initially decided they would adopt the GMAG guidelines. However, it was not possible in most other countries to easily introduce legislation to cover genetic manipulation because, unlike the United Kingdom, they did not have any appropriate existing laws that could be readily adapted. However, when the NIH guidelines were introduced and it subsequently became clear that they were not so demanding in containment requirements and had a comprehensive and codified system for categorization of experiments, almost all European countries decided to adopt the NIH guidelines. Another major factor was that national committees were also spared from having to spend considerable time, effort, and expense in producing their guidelines or codes of practice.

Implementation of the NIH guidelines in European countries varied widely, but in general these countries were more relaxed about surveillance of work going on than was the United Kingdom, which during the initial stages from 1977 to 1980 examined each proposed experiment on a case-by-case basis.

The national advisory committees that each country set up were really very similar to the GMAG in that appropriate government departments or learned scientific societies nominated eminent scientists to advise it, although it should be said they tended to be expert committees rather than the broad-based, representative committees such as the British GMAG.

Proposed European Community Legislation

The European Commission in 1977 recognized that there might be a need for a European directive for safety controls for rDNA work that would, if accepted by the European Parliament, be mandatory for the member countries. In their view, a directive was necessary to "harmonize" the controls so that there would be no unfair advantage, either gained or taken, within the countries of the European Community (EC). A draft prepared by the commission,[9] largely on the basis of Britain's GMAG procedures, was a comprehensive document covering not only rDNA work but also the use of organisms "created" by this technology. Even those who saw advantage in a directive criticized the draft legislation, particularly since it concerned the use of genetically manipulated organisms and the transfer of organisms from one center to another. The administration of such a program would have been formidable.

The European commission, on the advice of its Scientific and Technical

Research Committee, had proposed a "Multiannual Community Programme of Research and Development in Biomolecular Engineering." The original proposals for the program covered all aspects of laboratory work involving genetic engineering and, in addition to training scientists, included work on safety and risk assessment experiments. Any laboratory or center in the EC undertaking such work could apply to the commission, in competition with each other, for funds from the program. To avoid any conflict and problems relating to safety requirements which might arise between participating countries, the commission at that time (1978) believed it necessary to have a directive for rDNA work for this purpose. (The Multiannual Community Programme was approved in December 1981 at a total cost of £3.7 million, but has been radically pruned from the original proposals and is almost exclusively limited to agricultural work, training, and risk assessment studies.)

Since preparing the draft directive, there has been a considerable amount of new information available on safe host/vector systems and risk assessment procedures. To date, there has been no evidence to suggest that with all the work that has now been carried out either in Europe or the rest of the world has involved any novel or unknown hazards. The commission therefore, after seeking advice from experts and national representatives who had advised the European Science Foundation, prepared a draft recommendation,[10] which is essentially restricted to the registration of rDNA work to a national or local authority. A recommendation is not binding legislation on the member states whereas a directive is, and the member states must introduce the necessary national legislation to meet the requirements of the directive.

The draft recommendation was submitted on 4 August 1980 to the European Parliament for their consideration. The Parliament delayed discussion until the Economic and Social Committee (ECOSOC) of the EC gave its advice. The membership of ECOSOC includes representatives from the trades unions, industry, and the public nominated by the member states. Although the Economic and Social Committee is only advisory to the European Parliament, it can influence decisions that Parliament may take. Also, the commission may have to present a further case to the Parliament if ECOSOC's advice is in conflict with the commission's proposal. The ECOSOC called for a public debate (in May 1981) on the risks involved in rDNA work to decide what advice it should give to the European Parliament on the commission's recommendation rather than a directive. The ECOSOC published its report in October 1981[11] and recommended to the European Parliament that, in their view, a directive was the appropriate legal instrument to cover rDNA activities. However, the European Parliament approved the recommendation proposed by the commission on 22 February 1982. The Council of the European Community endorsed the recommendation in June 1982.

European Science Foundation

The European Science Foundation (ESF) created in 1974, is an international, nongovernmental organization designed to improve cooperation between European scientists and coordinate basic research. It provides a forum for member organizations to discuss common policies and joint activities.

In 1975, the Executive Council of the ESF set up an expert group to gather information on developments in the field of rDNA research. All European national rDNA committees were represented on the liaison committee as well as the European Molecular Biology Organization and the Commission of the European Communities. In addition, representatives from the NIH and the Canadian Medical Research Council have also attended meetings. The Liaison Committee was serviced by the Economic and Social Committee staff and Dr. J. Tooze from the European Molecular Biology Laboratory.

Although the ESF Liaison Committee has probably had little direct effort on either guidelines or research involving rDNA in Europe, it did provide a valuable forum for the exchange of information to keep national authorities apprised of progress in other countries. It was quite apparent during the years when guidelines were being formulated that misunderstandings or misinterpretation of information from national authorities gave rise to serious concerns among scientists—particularly those in the United Kingdom who thought that they were being unduly constrained. The ESF gave national authorities an opportunity to clarify through direct discussion the precise state of developments in other countries and thus allow their representatives to convey this information back to their own scientists.

At its meeting in January, 1981, the ESF Liaison Committee concluded that its task of harmonization of national guidelines was sufficiently complete and that it had fulfilled its mandate and should be disbanded. It was recognized that some countries were still reviewing their positions, but in general it was agreed that they would all reach a similar position in the near future. The Secretariat of the ESF has continued to circulate information on national guidelines to keep member organizations informed of progress. It was unanimously agreed that neither the laboratory work nor the industrial (large-scale) work involved any novel biohazard.

European Countries

The information in this section and the observations made are based on documents and discussions provided by the ESF Liaison Committee. Although the GMAG guidelines were initially widely accepted in Europe, they were very quickly dropped in favor of the NIH guidelines when it was realized that the NIH guidelines were less demanding than GMAG's. Since most academic research is conducted with government funds, it was not too difficult to ask that researchers follow the guidelines, although voluntarily. There was always the threat of financial sanction to those who did not observe the guidelines. The national advisory committees are therefore largely under the auspices of ministries or scientific/medical academies.

The Federal Republic of Germany

A centralized system of control exists in the Federal Republic of Germany and all centers doing rDNA work must register with the Zentrale Kommission für die Biologische Sicherheit (ZKBS). Local biosafety committees do not have the power to decide whether deviations are allowed from the national guidelines. The ZKBS issued its own guidelines based on the NIH scheme, but as with all codified guidelines or regulations they need constant revision to take into account new information. Therefore, there has always been a lag period before any modifications made by NIH have been implemented in Germany.

It is interesting to note that in Germany there is a very strong environmental lobby that has formed the basis of a political party, and strong demands have been made by them to formulate a new law to control rDNA work. Draft legislation was proposed in 1979 to appease their demands, but was never enacted by the Bundestag. This appears to be one of the reasons why the German lobby has been most active in pursuit of a European directive because, should a German law be introduced for rDNA, they will find themselves in the unique position of being the only EC country with a special law for this work. (The United Kingdom has regulations under the Health and Safety at Work Act.) Some pressure has also been applied by Germany in formulating the Scientific and Technical Committee biotechnology program by recommending that it be linked to a directive to control genetic manipulation work or at least have a substantial component set aside for risk assessment experiments.

The Netherlands

The Dutch guidelines are somewhat more restrictive than the NIH guidelines, and there remains a very real problem for scientists in Holland because local or regional safety committees can—and do—override the national Recombinant DNA Committee. For example, physical containment at the C-III level was forbidden. The Recombinant DNA Committee is, therefore, trying to negotiate with the government for statutory legislation to back the committee's advice and thus harmonize the situation within Holland and also with the rest of Europe.

The Dutch Committee in Charge of the Control of Genetic Manipulation (CCGM), under the auspices of the Royal Netherlands Academy of Arts and

Sciences, published in January 1981 its third and final report on guidelines, and its place has been taken by an ad hoc committee to assess containment levels for all projects—or at least until it considers that local institutional biosafety committees (IBCs) are competent to do so. The CCGM published its first report in 1977; its second report in 1979[12] recommended that the government allow at least one C-III containment laboratory. It also proposed that statutory regulations to control rDNA work should be promulgated.

In the third and final report (1979–1980),[13] the CCGM recommended that C-I and C-II research be permited, that C-III work be considered in one center and any C-IV work be done at the European Molecular Biology Laboratory facility in Heidelberg. The government has agreed to take over the advisory function carried out so far by the academy committee. Guidelines for rDNA work were introduced by the government in January 1981. All projects are assessed by the central government committee as it is considered that the local IBCs have not been sufficiently developed to delegate responsibility to them. The Dutch guidelines therefore at present remain the most restrictive in Europe.

Sweden

The Swedish regulations and procedure for the control of rDNA work are complex although in principle they accept the NIH guidelines. The National Advisory Committee is still being set up under the National Board of Occupational Safety and Health. Applications for new research have to be approved under the Occupational Health and Safety Act. It is, however, still a matter of debate what constitutes "new" research (new in Sweden or in the world?).

Other European Countries

Most other European countries are much more relaxed about the control of rDNA work, and they simply adopt the NIH guidelines and their revisions as they occur.

Effects of the rDNA Debate on the Support for Science in Europe

The U.K. Research Councils have given high priority to new imaginative initiatives in molecular biology, and when rDNA technology became available, support was given both directly to workers in councils' own units and indirectly through grants to university departments and research institutes.

The Medical Research Council in particular decided that in order to facilitate this work money would be provided for physical containment facilities. (This was, in the case of universities, a unique departure from the council's traditional policy of supporting only staff, equipment, and recurrent expenses in academic institutions.) Some £0.75 million has already been given for physical containment facilities, most of which are now redundant. The council also made it known that it would support the development of safe host/vector systems for use in rDNA work and also support work which would provide inforamtion for risk assessment or health care of those involved in the work. In any event, no significant demands were made on council because the containment requirements were progressively reduced to the point where most work is now done in Category I or under GMP. Support was also given for the development of a series of disabled *E. coli* strains and their subsequent testing in human volutneers.

The U.K. Health and Safety Executive, which has a small annual sum available for research, also gave, for a limited period, a modest sum to examine how effective containment prescribed for this work really was. This work was done at the Centre for Applied Microbiology and Research, Porton Down. The results contributed toward the final choice of figures used in the GMAG risk assessment procedure for the categorization scheme. However, the main value of the work has been in the contribution toward a greater understanding of the main organisms commonly used for investigation (i.e., *E. coli*, *B. subtilis*).

A Joint Working Party Report on Biotechnology (representing the Advisory Council for Applied Research and Development, Advisory Board for the Research Councils and The Royal Society) was issued in 1980[14], and recommended, among other things, that, the U.K. government should take greater initiative in coordinating biotechnology and providing further funds, the research councils should provide a minimum of £3 million annually for basic research, and the National Enterprise Board and National Research Development Corporation should set up a biotechnology company. The government has issued its response to the report (1981, Cmnd. 8177),[15] and despite its concern that private investment is much higher in the United States than in Europe and that firms in other countries may be seizing more quickly than U.K. companies the opportunities offered by biotechnology, the government's position is that an efficient and profitable industry can only be built by the private sector.

The setting up of a U.K. biotechnology company, Celltech, with funds provided jointly by private institutions and government was a unique departure for the United Kingdom. The Medical Research Council has a special agreement with Celltech which, if successful, should produce much needed additional funds for the council to support further basic research. Until this arrangement was agreed to by the government, any additional funds earned through patents or licenses were paid to the NRDC or the government and not necessarily used for further research.

It is doubtful that such commercial developments in government research organizations would have taken place without the recombinant DNA debate. Much attention to the exciting possibilities was focused by the debate, and this undoubtedly brought pressure to bear on the government to explore the industrial opportunities at an early stage—such as the setting up of Celltech.

The Medical Research Council has also set aside a number of training fellowships in biotechnology because it has been recognized there may well be a shortage of such staff in the immediate future. The aim of the fellowships is to enlarge the cadre of scientists capable of doing research on and teaching genetic manipulation in the United Kingdom. The fellowships are open to post-doctoral scientists, although a provisional limit has been set at ten fellowships, this could be increased if the demand is sufficient. In general, however, support for work in this area is still governed by peer review of excellence, and recombinant DNA projects have to compete for support with projects from other disciplines.

Commercialization of Recombinant DNA Research

The exciting commercial promise of this type of research has had an important influence on the political, financial, and regulatory aspects of rDNA work. It is perhaps worth emphasizing that large-scale work involves the use of a genetically manipulated organism and not genetic recombination per se. Although commercialism is well established in other academic fields such as engineering and chemistry, it is relatively new to biological research. Probably the commercial implications have become more important recently in terms of debate among scientists and politicians, rather than any novel hazards which may be associated with the work. It would appear that discussion of possible hazards associated with recombinant DNA work, particularly in the large-scale commercial situation, is only stimulated to achieve a political, commercial, or other objective such as the pressure created by some countries to have a European Directive control the work. Because the work has been highlighted in journals, through broadcast media, and in occasional parliamentary discussions, the speed with which the commercial interest has been aroused has probably been considerably hastened by all this publicity. This commercial pressure has led to tensions and divided loyalties among scientists, industries, and even governments. However, once the initial attraction of the commercial exploitation of the new technology has subsided, probably within the near future, then relationships may become a little less strained. The advantages of a closer cooperation between academia and industry, which has not previously existed so extensively in the biological field, may eventually become realized by scientists and lead to important advantages for the scientific community at large.

Some of the U.K. centers involved in the large-scale use of a genetically manipulated organism have generated wide local publicity to inform the public of their work, and no adverse comment or concern has been expressed at either the local or national level.

Possible Effects of the Recombinant DNA Debate on Future Technologies

No doubt confidence in the scientific community has been disturbed by the differences of opinion expressed by eminent and well-respected scientists. However, one suspects that this will not be long lasting, and already the impression is that discussions are less emotive and concentrate more on the potential of the technique rather than any damages that may ensue as a result of its use. The scientific profession on the whole are conscientious and sensible people, which is probably one reason why concern about recombinant DNA work was expressed by a group of scientists in the first place. Individual members of the scientific community have gone to considerable effort to explain the scientific background to the public and to the government committees. It is obviously important but extremely difficult to have some form of continuous updating—through public media—of scientific advances to keep the public informed.

Because of various coincidences in timing, the genuine concern publicly expressed by responsible scientists was used by some to make political and social disquiet to meet individual or organizational objectives. In particular, the concern surrounding health and safety at work, the jockeying for positions in the European Community programs, and the commercial jealousies intensified by the search for additional funds during a world-wide recession have all contributed to the scientific dilemma concerning the possible risks of rDNA work.

Sir Kenneth Clark once remarked that all great advances in civilization are based upon confidence. Scientists may have taken a setback in public confidence over the recombinant DNA debate, but they will almost certainly recover their position—although perhaps belatedly due in part to the reluctance of scientists to put their reasoned scientific case in the public arena. Responsible scientists will always try to present a fair case, but unless it is well reasoned and presented, the way will be left open for others to place difficulties and obstructions in the path of scientific work. This is especially true of the argument that *all* risks must be eliminated from, for example, working environments. The achievement of zero risk in any work, as demanded by some, is very difficult to attain, even apart from the cost involved, and, as with atomic energy, some scientists are beginning to present well-argued cases to set proper perspectives for the risk assessment of their work.

Even some government departments have seized on the opportunity to regulate and control more and more work—in some part to protect their interests in the event of increasing government cutbacks during the period of recession. The example one can quote here is the U.K. Health and Safety Commission consultative document issued in 1976, which defined work on the genetic manipulation of microorganisms so broadly that if it had been accepted it would have meant that all genetic experiments would have been encompassed under the legislation.

There is no doubt that many scientists and institutions remain concerned about what further controls and regulations may be placed on their work. There are other areas that at present remain unregulated—although there are known risks involved—such as work involving dangerous pathogens and carcinogens. It is interesting to note here that despite the lack of regulations in these areas, there are relatively few accidents. The safety record in work involving dangerous pathogens gives enormous credit to the scientific community at large in their ability to conduct their work in a safe and responsible manner.

Observations and Conclusions

The history of this controversy—or debate—suggests that scientists should analyze most carefully the impact that collective statements may have, particularly when issued by eminent scientists with the backing of a national government authority or respected scientific society. Then, those responsible for promulgating laws and regulations—especially in the sciences—should understand that they have a special responsibility, for the implications for science can be both financially and scientifically prohibitive. The introduction of flexibility into a legal framework to cater to scientific advances is very difficult indeed, and yet to continuously modify law or regulations when new information becomes available can be ridiculously time consuming.

There is no doubt that some organizations, having been primed by the recombinant DNA debate, will be on the alert for the opportunity to make new demands when another new technology comes along that they can turn to their own political ends, demanding involvement in the regulation or review of that particular activity. However, the recombinant DNA debate highlighted many moral issues that are only just beginning to be intensively discussed, and in the future there will be renewed efforts to introduce controls for this and other work for moral reasons.

Following the approval of the EC recommendation on genetic manipulation, member countries will be obliged to keep records of centers undertaking this work and also to ensure that adequate surveillance is provided so that no long-term problems arise as a result of the technology. Although one objective of the recommendation is to ensure harmonization of guidelines, it is difficult to envisage what these will be if radical changes occur in the near future. Most countries in Europe have now relaxed these guidelines so much that further relaxation might well mean abandoning them completely. Those countries that still retain stricter interpretation of the guidelines will no doubt relax them over the next few years to be in line with others.

Notes

1. *Report of the Working Party on the Experimental Manipulation of the Genetic Composition of Micro-Organisms* (*Ashby Working Party*), Cmnd. 5880, 1974, Her Majesty's Stationery Office, London.
2. *Report of the Working Party on the Practice of Genetic Manipulation* (*Williams Working Party*), Cmnd. 6600, 1976, Her Majesty's Stationery Office, London.
3. Dual funding arrangements are when the government, through the Department of Education and Science and the University Grants Committee, provides funds for individual universities for buildings, academic staff, recurrent expenditures, and some equipment. The universities are able to compete for research grants from the research councils that provide funds for additional staff, equipment, and recurrent costs of the research.
4. Members of this group included Professor W. F. Bodmer, Chairman, Sir David Evans, Professor D. C. Phillips, and Dr. W. G. P. Stocker.
5. "Flaws in GMAG's Guidelines," *Nature* 277 (1979): 509.
6. The Royal Society of London was established in the middle of the seventeenth century and first granted a Royal Charter in 1662. The main objective of the society is the advancement of natural knowledge.
7. Committee on Genetic Experimentation (COGENE) is organized by the International Council of Scientific Unions, an international nongovernmental scientific organization of autonomous international scientific unions.
8. *Third Report of the Genetic Manipulation Advisory Group*, Cmnd. 8665, 1982, Her Majesty's Stationery Office, London.
9. Commission of the European Communities, "Proposal for a Council Directive," COM(78)664, 1978.
10. Commission of the European Communities, "Draft Council Recommendation Concerning the Registration of Recombinant DNA Work," COM(80)467, 1980.
11. Economic and Social Committee of the European Committees, "Genetic Engineering: Safety of Recombinant DNA Work Colloquy," Brussels, October, 1981, Catalog No. ESC-81-014-EN.
12. *Second Report of the Committee in Charge of the Control on Genetic Manipulation*, Royal Netherlands Academy of Arts and Sciences, 1979, published by Korin Klyke, Nederlandse, Akademie van Weterschapper, Posbus 19121 1000GC Amsterdam.
13. *Third Report of the Royal Netherlands Academy of Arts and Sciences Committee in Charge of the Control on Genetic Manipulation*, 1979–80, published by Korin Klyke, Nederlandse, Akademie van Weterschapper, Posbus 19121 1000GC Amsterdam.
14. *Biotechnology: Report of a Joint Working Party*, 1980, Her Majesty's Stationery Office, London.
15. *Biotechnology*, Cmnd. 8177, 1981, Her Majesty's Stationery Office, London.

The concerns over recombining genes were first raised from within the scientific community. However, when the most strident objections to what was seen as a legitimate and rather benign research technique began coming from worried members of the public, including those with political power, the focus shifted. Then many scientists retrenched into a defensive posture, being careful not to question the safety of gene splicing, but rather to call attention to the technological wonders that awaited the human race. Some made lengthy and passionate public arguments defending the safety of cloning DNA in bacteria. Others tried to discredit the critics.

As the perceived threat of oppressive local ordinances and bureaucratic federal regulations intensified, so did the protests. But these tended to be rather in disarray, until the scientific professional organizations, most based in Washington, decided that it was necessary to organize, do their homework on federal law, and conduct themselves in a diplomatic manner. Otherwise, the gulf between the scientists and their critics would only continue to widen, with possibly disastrous results.

There were several organizations—the Association of American Medical Colleges, the Federation of Societies for Experimental Biology, and the American Society for Microbiology—that became active, both in educating their members in the ways of Washington and in telling Congress of the concerns of their constituents. Harlyn Halvorson, director of the Rosenstiel Basic Medical Sciences Research Center at Brandeis University and former president of the ASM, was one of the leading figures in the attempt to build a dialogue of reason between scientists and the Congress. If the ASM were ever to create the position of ambassador, Harlyn Halvorson would surely qualify as the first to hold that office.

5

The Impact of the Recombinant DNA Controversy on a Professional Scientific Society

HARLYN O. HALVORSON

One goal of this volume is to use the recombinant DNA debate as a case study to assist in the handling of other issues involving science policy. The American Society for Microbiology (ASM) was an active participant in the recombinant DNA controversy. This chapter does not represent the efforts of all scientists, or even of all microbiologists, but it does summarize a consensus effort on a major scientific policy issue.

A second goal is to describe how an issue of vital importance to a scientific discipline can influence its awareness about public policy and guide its professional organization in developing approaches. The recombinant DNA debate has indeed influenced the way in which ASM deals with public policy questions.

Early ASM Involvement in Public Affairs

Since 1899, the professional concerns of bacteriologists have been expressed primarily through ASM, an organization with a current membership of over 30,000. The ASM members work in the applied fields of infectious diseases, clinical microbiology, epidemiology, industrial fermentation, ecological microbiology, and food technology, as well as in the fundamental areas of immunology, genetics, virology, oncology, microbial physiology, environmental microbiology, mycology, and host-parasite interactions. By virtue of its size, scope, and services, ASM is a significant voice in the scientific community.

The recombinant DNA question coincided with a strengthening of ASM public affairs activities. Throughout its existence, ASM has been interested in and participated in scientific policy and has sent representatives to provide scientific advice to the government. It was particularly active during World War II. Since 1936, an ASM newsletter has informed the members on a variety of issues, and in 1956, the American Academy of Microbiology (AAM) was formed to enhance the professional affairs of microbiologists. For decades ASM has provided representation to bodies such as the National Research Council, National Safety Council, National Council on Health Laboratory Services, American Type Culture Collection, Internal Council of Scientific Unions, International Association of Microbiological Societies, and other bodies determining public policy.

The Council Policy Committee (CPC), which is analogous to a corporate board of directors, has long been aware of the need to make sure that congressional legislators and staff know the facts about the field of microbiology. ASM presidents have testified before congressional committees, and positions on important microbiological legislative issues have been communicated to appropriate House and Senate committees. Individual microbiologists were also encouraged to offer their best efforts to the shaping of government science objectives and action.

To fulfill this goal, an ad hoc committee was appointed in 1972 to study the participation of the ASM in public affairs. The committee recommended that ASM should have a public affairs committee and be involved in the public affairs arena:

> The concern of the committee should be with public affairs as they affect microbiology. The committee should articulate to the public and its representatives, in appropriate forums, the interests of microbiologists in legislation, ecology, public health, etc. The committee should recommend policy on present issues and should identify areas that may benefit in the future by statements of policy from the Society.

1973

In 1973, a number of events occurred that subsequently had profound effects on the role of ASM in public affairs. First, Dr. Carl Lamanna, long active in ASM, wrote to the members:

> The scientific community has a valuable and special experience to communicate to the making of public policy. . . . Predictably, if science is to contribute to the establishment of public policy it must speak through organized groups, namely, the societies of professional societies. . . . Legislators and most public

officials cannot be expected to be expert in scientific disciplines and academic pursuits. The ASM is one of the larger and more successful of scientific societies. Given the will to do so, it can develop the material needs to effect cooperation with other organizations in assuring a respectful hearing for the voice of science by the councils of government.[1]

Second, at its annual meeting in May, the Council of ASM authorized the formation of a permanent Public Affairs Committee (PAC) charged with developing appropriate mechanisms for recognition and response in areas of public policy that affect, or are affected by, microbiology. C.D. Cox of the University of Massachusetts was the first chairman.

Third, it was widely agreed that an effective and durable science policy would only be provided by a structure such as a reestablished Office of Science and Technology Policy (OSTP) headed by a science advisor appointed by the President of the United States. These concerns led ASM to join twelve other scientific societies in the formation of a Committee of Scientific Society Presidents (CSSP). This organization was largely responsible for reestablishing OSTP during the Ford administration.

Recombinant DNA and the Formation of PAC

At the first meeting of the newly created PAC in early 1974, there was considerable discussion of the new field of recombinant DNA research and possible consequences of this technology. Chairman Cox recommended that during the fall meeting of CPC there be a seminar to acquaint the officers of ASM with this new field and to discuss possible legislative consequences. On 25 November 1974, a seminar was held featuring Dr. Maxine Singer and Dr. L. Barron, both participants in the previous Gordon Conference on Nucleic Acids, and Dr. Ted Cooper, Assistant Secretary of Health, Education, and Welfare (HEW). This seminar not only helped to crystallize opinion within CPC and PAC, but also broke the news of the upcoming Asilomar meeting dealing with the implications of rDNA research. There was considerable concern over the fact that attendance at Asilomar was by invitation only. A further concern was the need to include expertise from medical microbiology. These concerns were expressed in a letter from Cox to Dr. Paul Berg,[2] one of the conference organizers, and as a result, Dr. Frank Young was accepted as a representative from ASM to the meeting. He subsequently reported to the membership and to PAC.[3]

Tooling Up

During the next year and a half, ASM was developing its capability for dealing with public affairs. A new executive director in 1974, with experience in governmental affairs and an appreciation of the role of microbiology in government, industry, and on the academic front, and a new public affairs officer were able to make a substantial impact through effective coordination of the public affairs activities of ASM.

With the approval of its Council Policy Committee, in 1974 ASM joined the Biology Alliance. This organization consisted of public affairs representatives from many biological societies, including the Federation of Societies for Experimental Biology (FASEB), the American Institute of Biological Sciences (AIBS), and the Association of American Medical Colleges (AAMC). The alliance provided a helpful liaison for collective efforts for those common issues affecting the member organizations. Meanwhile, PAC was preparing position papers on a variety of subjects, including national health legislation, energy, legislation regarding fetal research, regulation that would have restricted the transportation by airlines of etiologic agents, and genetic engineering.[4] At the same time, PAC was following the developments in the recombinant DNA controversy. Comments from ASM members began to accumulate,[5] and this led to two lead articles on the controversy in the May 1976 issue of *ASM News*.[6]

Endorsement of the Guidelines

During the spring of 1976, the Recombinant DNA Molecule Program Advisory Committee of the NIH was actively developing guidelines to regulate recombinant DNA research. This committee included three ASM members.[7] To acquaint the ASM membership with these activities Dr. DeWitt Stetton, Jr., deputy director for science at NIH, was invited to present the New Brunswick Lecture to the May meeting of ASM in Atlantic City; at that time he requested ASM endorsement of the NIH guidelines. He pointed out that NIH had no regulatory powers except by denial of research grants. He further asserted that moral persuasion would be most effective and that the assistance of the scientific societies was very important. This point was reviewed and deliberated by CPC during its meeting in Atlantic City and subsequently by the officers of ASM in early June. Since the NIH guidelines were not then available, it was decided to create an ad hoc committee to advise the CPC on what position ASM should take on the guidelines for biohazards. It was felt that our responsible position was to comment on the efficacy of the guidelines and containment facilities and not on the ethics of conducting research with recombinant molecules.

With the release of the guidelines by NIH on 23 June and subsequent publication in the *Federal Register* on 7 July, we were prepared to act. On behalf of CPC, as the then president of ASM, I appointed a review committee under the chairmanship of Dr. Harold S. Ginsberg, chairman of the Department of Microbiology, College of Physicians and Surgeons of Columbia University, to head this review of the guidelines.[8] The committee was designed to give a broad representation to areas involving the handling of pathogenic microorganisms, epidemiology, and medical microbiology, virology, and industry activities. None of the members had served in drawing up the initial guidelines.

On 11 November the ad hoc committee unanimously recommended endorsement of the guidelines with a few minor changes. They made two recommendations of their own: (1) ASM should assume leadership in the development of an educational program on the principles of contagion and host-parasite relationships for those engaged in recombinant DNA research, and (2) ASM should seek council from individuals qualified in evaluation of the adequacy and feasibility of the containment of plant systems. These recommendations were all promptly and unanimously approved by the CPC of the ASM and transmitted to NIH on 1 December 1976.

Into the Trenches

The Public Affairs Committee entered 1977 with a posture of monitoring activity on recombinant DNA. The first warning that events could change rapidly came with the legal report on recombinant DNA research by Dr. R. Riseberg, Office of the General Council, Public Health Division on 12 January. He concluded that "no single legal authority or combination of authorities exists which would clearly reach all such research and all requirements." There quickly followed the introduction of a number of restrictive bills to regulate this research (4 February, Senate bill S. 621 and 7 February, House bills H.R. 5020, 3591, and 92). We were alarmed. Dr. H. Whiteley, past president of ASM, and I readily accepted an invitation to an informal meeting on regulation by the director of NIH on 19 February. In response to a request for further comment, Dr. Whiteley wrote (7 March): "Advances in science are necessary for the health and well-being of our population. I believe that the general public as well as the scientific community should vigorously oppose legislation designed specifically to regulate research." I also responded to the director of NIH (8 March):

> We cannot effectively legislate against ignorance and accidents, regardless of facilities of licensing, and protection against such potentially hazardous research depends ultimately upon knowledge, judgment, and training in the use of hazardous microorganisms. We would support regulation containing licensing procedures so long as these were not so cumbersome as to impede the research itself and providing these regulations did not encompass existing research on infectious microorganisms not associated with recombinant DNA research. It should be remembered that a rather large number of microbiologists exist in this country who are actively and safely pursuing research on infectious, potentially hazardous microorganisms.

On 8 March Senator Howard Metzenbaum introduced a restrictive bill, S. 945, into the Senate, and on the following day Representative Paul G. Rogers introduced the administration bill, H.R. 6158, and his own bill, H.R. 4759. On 7 to 9 March, the National Academy of Sciences' forum on recombinant DNA took place in which a number of ASM members participated.

The stage was now set, and in the next few months ASM took an active role in the debate. Based on informal encounters, round tables during the annual meeting, responses to *ASM News,* PAC activities, and interactions related to congressional hearings, a task force was formed from the PAC, members of the Ginsberg review committee, officers of ASM, and outside ASM experts as needed. Based on the experience of previous years and in response to specific legislative initiatives, position papers were developed and reviewed by elective officers of ASM, AAMC, FASEB, AIBS, and others. We maintained by elective officers of ASM, AAMC, FASEB, AIBS, and others. We maintained close contact with NIH and testified before Representative Rogers's subcommittee in the House (17 March), before the Environmental Protection Agency (5 April), and before Senator Edward M. Kennedy's Subcommittee on Health and Scientific Research (5 April).[10] Our position was that scientific societies such as ASM had—and have—an obligation to speak out in the public interest on matters that require their scientific expertise for evaluating the issues. The recombinant DNA problem was fundamentally one of epidemiology, the science concerned with factors that influence the spread of diseases. Public discussions thus far had been carried out to a large extent by individuals ignorant of the principles of infectious diseases. We recommended informing the public so that it could exercise an intelligent influence on the technical problems involved in recombinant DNA research. The ASM, which included many members working on recombinant DNA, could provide expert advice on the problems involved in this new technology, for example, transmission of infectious diseases, knowledge of viruses, how to deal with epidemics, and knowledge and training in the safe use of hazardous microorganisms. Dr. Ginsberg, testifying before the EPA, concluded his testimony by stating, "To my mind in recombinant DNA research, the major biohazard is ignorance."

The next few weeks involved communication with the Rogers staff, with comments on H.R. 7418, and the Kennedy staff, with comments on S. 1217. By the first week in May (the time of the annual meeting of ASM), it was widely believed that federal legislation on recombinant DNA was imminent. Consequently, the PAC brought to CPC and the council of ASM nine points that were felt to be essential in the event federal legislation was evident. These were approved by the council on 9 May, transmitted to the membership meeting the next day, ultimately widely published,[11] and transmitted to Senator Kennedy and Congressman Rogers who were sponsoring legislation to regulate recombinant DNA research.

The following nine points were recommended:

1. That all responsibility for regulating action relative to the production and use of recombinant DNA molecules should be vested in HEW.
2. That to advise and assist the Secretary of HEW, an Advisory Committee should be established whose membership in addition to lay people should include representatives with appropriate technical expertise in this field.
3. That institutions and not individuals should be licensed.
4. That in each institution engaged in DNA recombinant activities, to the maximum extent possible, direct regulatory responsibility should be delegated to a local biohazard committee. These committees should include both members with appropriate expertise in the activities conducted at that institution, and representatives of the public.
5. That experiments requiring P1 requirements should be exempt from these regulations.
6. That license removal is an effective and sufficient deterrent to obtain compliance. Further, that ASM is opposed to the bonding of scientists or to the establishing of strict individual liability clauses in the conduct of DNA recombinant activities.
7. That ASM goes on record favoring uniform national standards governing DNA recombinant activities.
8. That the Secretary of HEW should have the flexibility to modify the regulations as further information becomes available. Further, we support the inclusion of a sunset clause in the legislation, i.e., that legislation will be reevaluated after a fixed period of time.
9. That ASM expresses its concern that in establishing such important legislation governing research, and that this proceed only after due and careful deliberation.

Developing a Strategy

By early 1977 it was clear to us at ASM and from advice from congressional professionals that to be effective and to maintain credibility on the broader issues of regulating research we needed to broaden the base of professional scientific societies.

ASM has not been alone in working on this problem. One of the most significant developments in 1977 was the creation of the Inter-Society Council for Biology and Medicine, an informal group of executive officers from seven biological societies, the Federation of American Societies of Experimental Biology, American Society of Allied Health Professions, Association of American Medical Colleges, American Institute of Biological Sciences, American Society for Medical Technology, National Society for Medical Research, and ASM. The council has enabled professional societies to act in concert on many issues and to delineate minor differences on others. The Inter-Society Council adopted the nine-point resolution and transmitted it to Representative Rogers. We then sought support from other scientific societies whose governing boards or elected officers ratified those nine principles. These organizations included FASEB, Genetics Society of America, American Genetics Association, AIBS, Biochemical Society, American Society for Cell Biology, CSSP, AAMC, Land Grant College Association, American Association of University Professors, American Society of Allied Health Professions, Tissue Culture Association, and many others. These organizations became part of a network to exchange information and to share views on pending legislation. At the same time a network of correspondents on recombinant DNA was formed within the ASM. This network enabled informed scientists to communicate their personal views on pending legislation to their respective senator or representative. Particularly during the summer of 1977 it became necessary to respond to many fire drills—to inform our officers, members, correspondents, and cooperating societies—on issues that required a response within a few days. Consequently, a great deal of time was spent developing a rapid communication network to inform, develop a consensus, and communicate this to the Congress and the administration. The unanimity of views by many organizations on basic principles not only made this task easier, particularly with small societies that do not have an office in Washington, D.C., but also reinforced our view that we were expressing a consensus position. At the same time we were involved with extensive discussions with members of the American Bar Association to review the legal aspects of regulation and with members of the administration, Office of Management and Budget (OMB), and OSTP, to explore their views.

Interactions with Congress

On 24 May 1977, Representative Rogers introduced his revised bill, H.R. 7418, incorporating many of the concerns of the coalition. It soon became clear that the less satisfactory status of legislation in the Senate had to be addressed. ASM responded on 27 May to Senator Kennedy with an analysis of S. 1217 and on 16 June supported Senator Jacob Javitz's amendment to S. 1217. On 21 June representatives of PAC and staff of ASM visited Senator Gaylord Nelson and two members of his staff. They were receptive to ASM views, which included (1) evidence of reduction in the risk associated with recombinant DNA experiments, and (2) the exemplary record of trained microbiologists in handling pathogenic microorgranisms and preventing the spread of laboratory-acquired infections.

On 18 July a group of scientists from six associations gathered in Washington for a visit with Senator Kennedy.[12] One of the main objectives of the meeting between the scientists and Kennedy was to convey to the Senator the nature of more recent scientific data that suggested that the risks associated with recombinant DNA research had been overstated. Kennedy listened attentively but did not seem to change his mind concerning his strongly held positions on (1) a national commission, (2) local preemption, and (3) monetary fines to be levied against individual scientists as punishment for regulatory infractions.

The meeting, however, did stimulate a number of articles in the *New York Times,* the *Washington Post,* the *Washington Star*, and *Science*. Indications from Capitol Hill suggested that Senator Kennedy and his staff were beginning to feel considerable pressure. Senators Gaylord Nelson, Alan Cranston, Thomas Eagleton, Adlai Stevenson, and others were noticeably shifting away from the Kennedy position and toward more moderate legislation.

On 2 August 1977 a significant change occurred. On that date, Senator Nelson presented amendment no. 754 to the Recombinant DNA Act, S. 1217. The amendment was a substitute to S. 1217 and reflected new information and views. It differed from Senate bill S. 1217 and from H.R. 7897, the latter then pending before the House Interstate and Foreign Commerce Committee, with respect to a number of major issues: the nature and extent of regulation necessary to combat potential risks from DNA research activities, the definition of recombinant DNA activities to be regulated, penalties for compliance, and the authority of the federal government to preempt state and local regulations.[13]

The Congress recessed in August without acting on the legislation. During the late summer recess the coalition was active in its support of the Nelson amendment. By the time Congress reassembled in late September, there were many cosponsors of the Nelson amendment and, from our poll of the Senate, a majority support.

Senator Kennedy, responding to information indicating that no harmful effects or spread of recombinant organisms had been reported since recombinant DNA technology had been introduced four years earlier, told the Association of Medical Writers (in a telephone speech to their meeting in New York City on 27 September) that he was planning to introduce a new recombinant DNA bill, a simple extension of the current NIH guidelines to include all individuals and organizations involved in recombinant DNA research that would have a one-year sunset clause. Kennedy said he would also create a study commission, with public and scientific members, to explore issues in depth and make recommendations for future action.

Following Kennedy's speech to the medical writers, a member of his staff contacted me and asked if we would review and comment on the proposed new bill. We agreed to study the new proposal in great detail and compare it with other legislative proposals in the hope that a "conferenceable" bill would emerge from the Senate.[14] On 13 October I wrote to Senator Kennedy on behalf of the Inter-Society Council for Biology and Medicine, noting that he had proposed in his speech on 27 September "to extend the NIH guidelines to all parties conducting recombinant DNA research, [to urge] the Congress to enact this new bill and to defer permanent legislation for one year, [and] to form a national recombinant DNA study commission." I pointed out that the three-page draft bill did not allude to a time limit of one year, did not make reference to the establishment of a national study commission, and that, since it was unclear how the differences between the bill and Senator Kennedy's public statements were to be reconciled, the Inter-Society Council of Biology and Medicine and other members of the scientific community were reluctant to comment on the draft before them. I also noted that the draft bill made no reference to administrative arrangements, assignment of responsibility for administering the law, or a legal basis for addressing such controversial questions as federal versus local preemption and disclosure of information. We did not and, to date, have not received a response to this letter.

Meanwhile, we were encouraged by Senator Adlai Stevenson's statement in the 27 September *Congressional Record*[15] on the state of the art regarding recombinant DNA legislation and by a letter he sent to Frank Press, science advisor to the President. In his letter to Press, Stevenson called most of the present legislation "ill-designed" to achieve the objective sought, namely, protection of the public without impeding research. He said he was concerned about government interference in freedom of scientific inquiry and that his Subcommittee on Science, Technology, and Space was seriously considering oversight hearings to provide a forum for receiving up-to-date assessments of risks associated with recombinant DNA and suggestions for the proper government role in light of the new assessments. Stevenson said he would also explore existing statutes to regulate recombinant DNA research. Calling for further deliberations by Congress before enacting legislation, Stevenson asked Press for his comments on whether the administration shared his concerns.

Under the reorganization of the Senate committee jurisdictions, Stevenson's subcommittee had both legislative and oversight authority over science, engineering, and technology research and development and policy, and hearings were held on 2, 8, and 10 November. ASM participated in these hearings and summarized the positions developed with the coalition. Briefly, these were concerns about the apparent intemperate rush to establish legislation to regulate recombinant research without first consulting with the appropriately qualified scientific and medical experts, the need to understand that early allegations concerning recombinant DNA research were characterized by uncontrolled imagination and excessive claims by individuals who lacked knowledge of infectious disease, and the need for minimal interim legislation to extend appropriate guidelines to all recombinant DNA activities regardless of funding source.

We pointed out again that there was a steady proliferation of local community activity to limit, impede, or halt DNA research, making federal preemption over local laws desirable. Otherwise, there would be a patchwork of conflicting laws regulating microorganisms that recognized no political or geographical boundary. We stated again that the protection of the public against potentially hazardous microorganisms depends ultimately upon knowledge, judgment, and training in the study of pathogenic microorganisms and that the ASM had committed itself to providing workshops and assistance in training and certification in the areas of infectious disease and medical microbiology.

Our concern about training was taken seriously by the director of NIH and also by the directors of the National Institute of Allergy and Infectious Diseases (NIAID), the National Institute of General Medical Sciences, and the National Cancer Institute. NIAID provided support for ASM to establish a working panel to develop minimum standards for training in microbiological techniques and biohazard control for participants in recombinant DNA research.[16]

The working panel met first on 16 and 17 December 1977 in Washington, D.C., and had a draft ready for comment by fall 1978. Early in 1979, a report went to NIH. It was subsequently to become a recommended training program by NIH implemented through NIH support to the University of Minnesota and ASM.

During the fall of 1977 and in 1978, we continued to work closely with the Rogers committee and the Science and Technology Committee on further legislative efforts in the House. Preemption continued to be one of the main concerns. During this period, a number of ASM members and practitioners of recombinant DNA research urged us to drop the preemption issue, since the prospect of federal legislation had declined. This led to a reappraisal within the coalition and reaffirmation of our earlier stand.

DNA and the Law

During the period of developing position papers, ASM had extensive contact with members of the American Bar Association, members of law faculties, the American Civil Liberties Union, the Environmental Defense Fund, and various insurance companies. We sought advice on first amendment rights, effect of penalties in various legislative bills, preemption, protection against lawsuits, patentability, regulation of the private sector, and restraints on basic science. We accepted all opportunities to explain our position to state legislators, their committees, and local governing bodies. ASM members participated in the initial ordinance hearings in Cambridge, Massachusetts, and in similar hearings across the country, pointing out that if a public health problem existed from this research it was a national, not a local, problem.

Ultimately, ASM provided *amicus curiae* briefs in two court cases. The first involved a suit in May 1977. Ferdinand J. Mack filled a suit in the U.S. District Court of Appeals on behalf of his two-and-a-half-year-old son requesting an injunction against the NIH scientist conducting certain recombinant DNA experiments at the Frederick Cancer Research Center at Fort Detrick, Maryland. Mack charged that unless a preliminary injunction were granted, the experiments would cause irreparable injury. He contended that because of the nature of the organism to be created by the research, even a minuscule quantity, if released into the environment, could pose a threat to life and health. He also charged that the environmental impact statement prepared by the NIH did not comply with the requirements of the National Environmental Policy Act (NEPA).

A temporary restraining order was issued by the court at the time of Mack's suit restraining NIH scientists from conducting risk assessment experiments to test the biological properties of polyoma virus cloned in bacterial cells. The experiments were intended to test certain worst-case scenarios by putting polyoma virus DNA into bacteria and then tracing whether they could get into the cells of mice infected with those bacteria. ASM argued that the degree of hazard should be subject to experimental tests to determine if there was indeed a risk, and that these tests should be carried out under rigid containment conditions and the research compared to our historically safe experience in dealing with laboratory-acquired infections—fewer than 4000 cases reported in the last century.

The court referred to the brief filed by ASM, acting as *amicus curiae* for the NIH, which said that the weight of scientific opinion now considered that recombinant DNA research in accordance with the NIH guidelines would not have an adverse environmental or public health consequence.[17]

In denying the injunction, U.S. District Court Judge John L. Smith, Jr., cited a 1972 Supreme Court decision saying that "the only role for a court is to insure that the agency has taken a hard look at environmental consequences." He ruled that the environmental impact statement did represent a "hard look" by NIH at recombinant DNA research performed in accordance with its guidelines. It appeared that compliance with the NIH guidelines would insure that no recombinant DNA molecules would escape from the carefully controlled laboratory into the environment. The judge said that NIH had carefully considered the potential risks of these experiments under the guidelines and had taken the necessary precautions; the plaintiff, however, "has not shown that we would be irreparably injured." Accordingly, the plaintiff's motion for a preliminary injunction was denied.

The second instance, an *amicus curiae* brief to the Supreme Court, involved a patent claim. On 16 June 1980, the Supreme Court ruled by a 5 to 4 decision in the case of *Diamond* V. *Chakrabarty* that Dr. Chakrabarty's oil-eating bacteria, produced by a shuffling of their DNA plasmids, were living inventions. This was a dramatic and important decision which has profound effects on the broader issue of patenting life forms. Since World War II, microbiological patent law has evolved on a case-by-case basis. The court decided in the Chakrabarty case that a product of genetic engineering resulting from a physical manipulation by man to produce a new microorganism is a product which is a "manufacture" or "composition of matter" within the meaning of the current patent law. The patenting of such "living inventions," however, leaves many problems unresolved.

Seven Years Later: Where Have All the Monsters Gone?

What are current attitudes of ASM on recombinant DNA? Research experience and targeted risk assessment experiments have shown no overt damage from this technology to either personnel or the environment.[18] Also, experiments with organisms containing recombinant DNA inserts show that such organisms are at a competitive disadvantage. Further the discovery of spliced genes in eukaryotic cells greatly reduces their ability to be expressed in shotgun experiments in bacterial cells. Thus recent research and our collective experience has clearly shown that many of the assumptions underlying initial concerns over this technology were unwarranted. ASM continues to believe that recombinant DNA is not only a safe but also a powerful tool to analyze and understand the regulation of the chromosome itself. This may be its most important contribution. In addition, recombinant DNA technology is leading to the production of valuable human proteins, in formerly unavailable quantities, which will significantly improve therapy for certain medical conditions. This is not to say that all concerns over recombinant DNA are gone, and ASM favored uniform application of the NIH guidelines to all recombinant DNA research regardless of funding.

The new biotechnology, which has grown out of financial support from NIH, the National Science Foundation (NSF), and other federal agencies, brings with it new concerns. No one would argue the need for a company to recoup expenses and profit from its risks. At the same time, it is prudent to ask whether such patenting is in the public interest. At one level, much of the basic biological research has been supported from public funds (usually NIH and NSF). Now, however, biology is entering the same arena of endeavors as chemistry and engineering, in which both public and private funds provide support. For several decades these two fields have struggled with the problems of confidentiality, priority rights, and the academic freedom of open inquiry. A review of this history provides no clear lessons for recombinant DNA biology but does indicate that fundamental public policy questions remain to be resolved and that conflicts of interest will arise in the next few years. These problems are already troubling some of our universities as they struggle with new ways to raise income.

ASM has explored the issue of patentability in cooperation with the House Committee on Science and Technology in a forum on patentability of microorganisms.[19]

Together We Stand

Initially, ASM was motivated by the concern that its expertise should be involved in any public issue in which microbiology was a prominent feature. In view of the fact that any legal decisions would have a severe impact on the discipline, the society felt it should carefully analyze and respond to the effect on its professional activities. In a broad sense this has been a continuation of the concern that professional societies should take both an interest and responsibility in matters in which their expertise is involved. As the recombinant DNA debate developed, it quickly became clear that it involved responsibilities far beyond those of microbiology itself: regulation of research, risk-standard evaluation, and public confidence. These considerations led ASM to seek a broader consensus among other professional societies who share a common concern. ASM sought to have this consensus based on the vote of the other societies' governing bodies and, where possible, of its own membership.

Many other issues soon arose. A number of activities and organizational coalitions and/or cooperations were formed to respond to these from either congressional or regulatory initiatives.

Formation of the Board of Public and Scientific Affairs

Over two decades ago ASM relocated from Ann Arbor, Michigan, to Washington, D.C., to represent microbiology more effectively. In the years since Asilomar, many changes have occurred the American Society for Microbiology. With the formation of PAC in 1974, the task force concept was introduced. As each major issue arose, a small group of experts was assembled as a task force to advise PAC, headquarters, and ASM members on matters of substance. From such contributions an ASM position could be formulated rather quickly. The task forces were in essence ad hoc committees, and they have been extremely effective on a number of biomedical issues. Some problems were handled by expert advice from standing ASM and AAM committees, but often the problem arose suddenly, with an ASM position requiring expert advice quickly obtained by telephone or a short meeting via the task force concept.

The evolution of this program in ASM led to the creation of a Board of Public and Scientific Affairs with a number of subcommittees to enable the society to broaden the base of scientists involved and to provide a longer period of evaluation prior to public response. We have been trying to move away from fire drills to a situation in which policy positions can be deliberated, reviewed, approved by officers, and transmitted first to our membership. Time does not always permit this, but to the extent possible we attempt to predict our problems. Public policy sessions at our annual meetings, workshops, involvement of other boards such as those of Education and

Training, the American Academy of Microbiology, and publications complement briefing sessions with our elected officers and CPC. All this has led to better communications through newsletters, correspondence, cooperating societies, and through our Congressional Fellowship Program, to Congress. Issues have been broadened to include ethical questions, regulations, and particularly education and training. These more sophisticated approaches have been ones of implication rather than shifts in conception.

Planning for the Future

A number of lessons from our experience with the recombinant DNA debate may provide guidance for the future. I believe ASM responded appropriately. Although we started somewhat unprepared and had to develop our response system rapidly, we sought specialists in our field and used their expertise to form a broad consensus. On the basis of this we sought to respond and inform the executive, legislative, and judicial branches of government and the public. We still believe that this is the proper role for a professional society, as long as it speaks on the basis of its scientific expertise, provides balanced opportunity for the scientific issues to be debated, and assumes responsibility for education and training. Unfortunately, we encounter a malaise that comes from the public's excessive expectations concerning the ability of institutions to respond with a balanced analysis. Scientists are responsible for supplying the public and its representatives with all relevant facts and uncertainties in a responsible manner. We should give serious thought to the possibility that our existing scientific institutions are incapable of evaluating the facts, accurately presenting them to the public, and of being fully aware of their moral obligations. Ultimately, the public must decide for itself what risks (economic, environmental, and so forth) will be tolerated to achieve expected benefits. But if the public is to participate knowledgeably, it needs a clear statement of the issues involved and an evaluation by technical experts of the potential benefits and risks. This I see as our challenge for the 1980s and beyond.

Problems of Scientific Societies Interacting in the Public Arena

The experience of ASM in the recombinant DNA debate highlighted a number of problems undoubtedly shared by other societies dealing with public affairs. These are:

1. How do societies obtain timely responses? The response time is often short, the issues are complex, and the meetings of the societies infrequent. To the extent possible societies must anticipate and articulate future problems.

2. How do societies determine their positions? There is a need to inform the constituents of the society and obtain prompt feedback on the issues. After a position has been defined, it must be officially endorsed by the officers and the society itself.
3. How do societies avoid the charge of being lobbyists? First, they must be sensitive to the problems of self-serving activities. Societies should develop well-reasoned positions based on their expertise, publish these for comment, and provide opportunities for differing views to be expressed.
4. How do societies educate their own members? There are several responsibilities societies have to address. During debates on issues societies should attempt to provide their members with a clear statement of the risks. Members need to be informed on new technologies and the safety factors related to them. Finally, societies have an obligation to provide training for new technologies.

Lessons Learned

There were several lessons ASM learned from working with a new technology. Since the recombinant DNA debate was a forerunner of problems yet to come, it is important that we evaluate our own lessons early.

First, we gained experience and confidence in the way to approach public policy issues. Those issues clearly identified with our expertise were chosen, the problem defined, and through interactions with our committees and members a position was developed. This was given to our elected officials for approval, and where possible, CPC and Council.

Second, we learned much about the problem of how a nation gets technical groups to provide advice in times of controversy. Our effort was to develop coalitions in a search for broad mechanisms for input necessary in our pluralistic society. We need to search for better and more effective mechanisms.

Finally, there is a need to provide technical advice to the legislative, executive, and judiciary branches on many problems involving science and technology. We support, and shall continue to support, the Congressional Research Service, the Congressional Office of Technology Assessment, OSTP, and the Congressional Fellows Program. However, mechanisms have yet to be developed to provide advice to the judiciary.

Acknowledgments

The involvement of scientific societies in the debate involved thousands of individuals, far too many to list. But in ASM I would especially like to acknowledge the help of Robert Acker, our former executive director; Don Cox, PAC chairman; Robert Watkins and Janet Shoemaker, PAC staff; Helen Whitely, Fred Rasmussen, Ed Lennette, Brinton Miller, and Joe Joseph, elected officers; and members Frank Young, Harold Ginsberg, Roy Curtiss, Bernard Davis, W. Rowe, and Maxine Singer. Also we are indebted to Walt Ellis, Gene Hess, and Robert Bock (FASEB); John Sherman and Tom Morgan (AAMC); Robert Trumbull (AIBS); Ernest Gilmont (CSSP); and Walter Milne (MIT), as well as to dedicated congressional staff such as Burke Zimmerman, who eased our interpretive task.

Notes

1. C. Lamanna, "Lobbying and the Professional Scientific Society," *ASM News* 39 (1973): 225–26.
2. Letter from C.D. Cox to Paul Berg, 3 December 1974:

    As we understood from our recent meeting, there will be a session held in February at Asilomar to review research in this area and to develop public policy recommendations. As we understand the nature of this meeting, it will deal not only with plasmid biology, but also the joining of oncogenic viruses and eukaryotic DNA fragments to bacterial plasmids and their implications. We feel that this general topic embraces a number of areas in which we are not only concerned, but in which the Society has had a great deal of relevant experience and has an important voice to bear with regard to public policy matters. In particular, we recognize that the transfer of resistant markers from one bacterial strain to another can involve important problems of medical microbiology and influence both our activities with regard to public policy and with regard to a long-standing area of recommendations for legislation for microbiology. We further understand that likely to arise from your meeting in Asilomar and subsequent decisions will be recommendations for the handling of potentially dangerous microbial biohazards.

    I would like to point out that the Society, through both its training and [its] recommendations for handling of microorganisms, has had a long experience with pathogenic organisms and with concern in problems in virology. In addition, another segment of our Society has concerned itself with the selection and maintenance of strains which are resistant to various antibiotics. Consequently, we feel that through the expertise in the Society we can bring to bear on an important problem raised by you and your colleagues information and concerns over training which are relevant to policy decision matters. We would like an opportunity to be represented at this critical meeting in an area in which we feel we have a substantive input and major stake."
3. F.R. Young, "Report from Asilomar," *ASM News* 41 (1975): 260–61.
4. As articulated by Chairman Cox, the two main tasks of PAC were "(1) monitoring the congressional and federal scene and providing information to the ASM membership, and (2) mustering expert information and advice from the ASM membership for appropriate testimony before committees of Congress and vari-

ous federal agencies." See C.D. Cox, "Public Affairs Activation of ASM," *ASM News* 42 (1976): 216–18.

5. B.D. Davis, "Genetic Engineering: How Great Is the Danger?" *Science* 186 (1974): 309; I.P. Crawford, "Recombinant DNA Molecules," letter to the editor, *Genetics* 78 (October 1974); J.K. Bhattacharjee, "Recombinant DNA Molecules," *ASM News* 41 (1975): 445.
6. P. Berg, "Genetic Engineering: Challenge and Responsibility," *ASM News* 42 (1976): 273–77; M.F. Singer, "Summary of the Proposed Guidelines," *ASM News* 42 (1976): 277–87.
7. The three ASM members on the NIH committee were Roy Curtiss III, professor of microbiology, University of Alabama, Birmingham, Alabama Medical Center; Waclaw Szybalski, professor of oncology, University of Wisconsin, Madison; and Jane Setlow, molecular biologist, Brookhaven National Laboratory.
8. The ASM review committee members included Dr. Carl Johnson, formerly of the Centers for Disease Control, Atlanta, Georgia; Dr. Robert Huffaker, associate director of the Office of Public Health, New York State Health Department, Albany; Dr. Jerome Birnbaum, vice president of microbiology and agricultural research, Merck, Sharp and Dohme; Dr. Frank Young, dean of the School of Medicine and Dentistry and vice president for health affairs, University of Rochester; and Dr. Curtis Thorne, professor of microbiology, University of Massachusetts, Amherst.
9. Memorandum to director of NIH, 12 January 1977.
10. R. Watkins, "ASM Representatives Express Views on Recombinant DNA," *ASM News* 43 (1977): 249–51.
11. H.O. Halvorson, "ASM on Recombinant DNA," *Science* 196 (1977): 1154; and Halvorson, "Recombinant DNA," *Chem. Eng. News* 55 (1977): 4.
12. The scientists were H.O. Halvorson, ASM; Lawrence Bogorad, Harvard University (FASEB); Peter R. Day, University of Connecticut (AIBS); Oliver Smithies, University of Wisconsin (ASBC, FASEB); Tracy A. Sonneborn, Indiana University (Genetics Society of America); and Frank E. Young, University of Rochester (AAMC).
13. The proposal reflected concerns expressed in supplemental views filed with the report (no. 95-359) of the Human Resources Committee on S. 1217.
14. Letter from the author to Senator Edward Kennedy, 4 October 1977:

    As you know, ASM has been working closely with the Inter-Society Council for Biology and Medicine and other scientific and professional societies in formulating responsible and thoughtful responses to the congressional bills and amendments to regulate recombinant DNA activities. The effort to evaluate and respond to the bills has been a collective one, shared by many persons and organizations. In order to provide you with careful review of this new proposal, members of the coalition must be consulted and given adequate time to study and comment on the bill.

    Many members of the coalition have not seen the new bill, but have expressed an interest in the possibilities they understand it offers. We wish to thank you for reconsidering S. 1217 and for drafting a new bill. We will study this interesting new proposal in great detail and compare it with other legislative proposals. We sincerely hope that a "conferenceable" bill will emerge from the Senate.

15. A.E. Stevenson, "Recombinant DNA Legislation," *Congressional Record,* Senate, 1977, S15410–S15413.
16. The contract was awarded on 14 September 1977 with the following technical specifications:

    "*Independently,* and not as an agent of the Government, the contractor shall exert its best efforts to conduct the necessary activities to develop a set of minimum standards for training in microbiological techniques for participants in recombinant DNA research under NIH Guidelines relating to experiments requiring P1 through P3 containment conditions."
17. Brief of the American Society for Microbiology Amicus Curiae *Mack* vs. *Califano* 447 F. Suppl. D.D.C., 1978.
18. Susan Gottesman, "Report on Workshop on Recombinant DNA Risk Assessment," *Recombinant DNA Technical Bulletin* 3, no. 3 (November 1980): 118; and S.L. Gorbach, "Recombinant DNA, An Infectious Disease Perspective," *Journal of Infectious Diseases* 137 (May 1978): 615–23.
19. Robert F. Acker and I. Schaechter, *Patentability of Microorganisms,* ASM Publications, 1981, a reprint of proceedings from the July 1980 forum on the patentability of microorganisms, cosponsored by the American Society for Microbiology and the Committee of Science and Technology, U.S. House of Representatives.

The extent to which any event or issue is communicated to the public, and the manner in which it is done, is perhaps the most important determinant of the level of interest or concern in that issue. Even so-called objective reporting, as we are led to believe we see on the evening network news, can influence viewers in many different directions, depending upon the selection of material and specific points of focus. For example, in the late 1960s and early 1970s the news coverage of the several demonstrations in Washington protesting American actions in Southeast Asia always managed to single out and focus upon the various radical fringe elements participating and their bizarre activities, rather than the fact that the demonstrations drew huge numbers of ordinary citizens speaking their conscience. Those whose only source of information was their television set came away with a distorted impression of what those demonstrations were all about.

In a country where news is big business, it is no wonder that the focus is always on the melodramatic, the lurid, or the ominous, if it is possible to squeeze such emotion-provoking aspects out of apparently ordinary goings on. The advent of gene splicing provided both the paper and electronic media with plenty of copy, not necessarily real events presented in clear perspective, but subjects which allowed fertile imaginations to paint all sorts of scary pictures of our genetically manipulated future.

Still, there were capable and responsible science writers who could make the often obtuse machinations of science and government both interesting and accurate, and still compete with their less restrained colleagues. Harold Schmeck, Jr., with the *New York Times* for many years, was a witness to the recombinant DNA controversy from the beginning. His coverage of this often difficult and fractious affair set a standard for clarity and objectivity. But his competitors were many, and often attracted more attention with their more colorful, if less accurate, accounts. Here, Harold Schmeck turns his attention to the media itself and the role it played in directing the course of the recombinant DNA controversy.

6

Recombinant DNA Controversy: The Right to Know— And to Worry

HAROLD SCHMECK, JR.

The most important single thing about the recombinant DNA controversy may have been the way it was perceived by the public. In a sense it was all an exercise in human imagination. The controversy rested on no tangible, identifiable event such as fire, famine, war, or pestilence. Not a single person was killed, injured, or even made ill. On the other hand no one was cured of anything or, for the time being, enriched in any way. Yet a seething, nationwide controversy erupted. The aftershocks have clearly affected the national consciousness and some important national policies and are still reverberating today.

The available evidence continues to indicate that no one has been injured by any gene-splicing research, nor has anyone been cured of anything by its applications, although some entrepreneurs are anticipating fortunes from the future possibilities of the field.

None of this implies that the whole issue was illusory, trivial, or a fake. The concern was legitimate, and was raised not by chronic alarmists, but by some of the scientists most involved in the research. In general, their most serious original fears proved to be unfounded, although there is no guarantee that the hazards that concerned them will always remain hypothetical, particularly now that industrial applications and a much accelerated pace of research are broadening the scope and multiplying the amount of the work.

But, demonstrably, the public's perception of the possible risks and benefits of recombinant DNA work did have an impact on the research, continues to have an impact today, and, in the future, could have a profound influence on the progress of scientific research, industrial development, and society's safety.

Widespread public hysteria after the Asilomar conference might have resulted in a forced halt on all the research with recombinant DNA techniques. This might have lasted for years. Under those circumstances the work would simply have gone underground or abroad, with profound ill-effects on American science and its usefulness to society at large.

Today, too much public apathy might encourage irresponsible use of the new techniques with all manner of bad consequences. For those reasons alone, scientists are not justified in thinking the long debate and its fallout in the form of guidelines, congressional hearings, and related matters were of no consequence. Nor is there any question that the public has every right and every need to know and understand the issues exemplified by the recombinant DNA controversy.

Given those circumstances it may be worthwhile to consider the role of the press in these events with some attention to why writers and editors do some of the things they do and why they do not do others. No attempt will be made here to assess the overall quality or effects of news coverage of recombinant DNA technology. To do so would require large-scale and lengthy research. So far as I can determine, no project of that scope and quality has yet been published anywhere by anyone.

The problem is that of providing the necessary information to the public. This cannot be left entirely to the scientists, nor to the pressure groups that seemed to spring up like mushrooms on all sides of the controversy. These many sources of information and opinion need to be heard, but the main vehicle through which they will get to the public will certainly be the press and its electronic counterparts. These so-called media of communications are an imperfect means for telling an imperfect audience about their imperfect world. But, for all the imperfections, the task and its objectives are vitally important.

Because the controversy itself has had tangible effects and because something like it is likely to happen again with a different cast of characters and a different set of worries, the history is worth some reflection.

From the point of view of public knowledge, the recombinant DNA controversy began almost a year late. It should actually have become a public issue in the early fall of 1973 as soon as the letter signed by Dr. Maxine Singer and Dr. Dieter Söll appeared in the 21 September issue of *Science*.

Any statement from a reputable source that begins: "We are writing to you, on behalf of a number of scientists, to communicate a matter of deep concern . . ." deserves attention from the press.[1] In fact it received very little attention. The most notable exception was an article by Nicholas Wade in the News and Comment section of *Science* shortly after the letter was published in that journal. But even though it has a large circulation, *Science* is not generally a source of news for the general public. And the item in *Science* was apparently missed by most of the general press.

In any event the glad tidings that some biologists feared they were about to risk ruining three billion years of evolution or at least start a worldwide plague or two did not impress itself on the public consciousness until just before the letter from Paul Berg and his colleagues in July 1974.[2] By that time the issue had been percolating through the scientific community long enough to have reached at least the more alert segments of the press.

One of the writers for the general press who had learned about it earliest was Victor McElheny of the *New York Times*. Indeed, he may have been the first, but that would be difficult to document. He had become possessed by the thought that the *Times* was not devoting enough coverage to molecular biology and had set out to rectify this deficiency. In the pursuit of scientific developments in this rapidly advancing field, he, almost inevitably, encountered the controversy. He was therefore already planning a story on the subject before the Berg letter was published.

At least two other science reporters, Stuart Auerbach of the *Washington Post* and David Perlman of the *San Francisco Chronicle*, had also learned about it.

At this point a story that had been lying fallow, so to speak, for almost a year, abruptly became highly competitive to journals and correspondingly important. It was clearly significant of itself and further important because it had become a competitive news story. This point should be reasonably familiar to scientists, among whom, I believe, there is also a certain spirit of competition for priority of publication.

The circumstance may be worth dwelling on for a moment. The news at hand was no trivial matter. A group of specialists in an important realm of biological science wanted a moratorium on some experiments in their field because they were concerned for the public safety. They were sufficiently concerned to ask three of the most important scientific journals (*Science, Proceedings of the National Academy of Sciences,* and *Nature*) to publish their letter simultaneously here and abroad, and the National Academy of Sciences called a press conference to explain it all.

There may have been a precedent for such a set of events, but no one could readily recall one. The subject seemed clearly related to then-current concerns over environmental pollution. It also seemed to have implications for the sanctity of human and animal genes, a subject full of implications for something very important indeed—the rights and freedom of the individual. All of these factors were calculated to make it big news.

So when the story did finally come out, it had been in a pressure cooker of largely internal debate in the scientific community for more than a year. Then it blew up suddenly under intrinsically alarming circumstances with major elements of the press fighting to avoid being left behind.

In short, the news was finally put out under circumstances that plainly invited outside spectators to conclude that the cataclysm was at hand. The message could hardly have been put more forcefully if the timing and manner of publication had been deliberately incendiary. The fact that neither the scientists involved nor the reporters really meant to broadcast a message of clear and present danger is irrelevant. And that too may be worth remembering. The press conference, incidentally, was an entirely sensible idea, despite the reported objections of some of the scientists involved. To have put out such a message without any effort to explain it would have made matters much worse than they actually were.

All of this, of course, is an egregious indulgence in hindsight, but that is, after all, one of the themes of this book.

It is at least worth speculating that the outcome might have been different if the news had become general in September 1973. Then the precipitating event would have been one letter in *Science* from a small, even though important, group of specialists. The news would still have been alarming to many, probably justifiably so, but it would have lacked many of the artificial ingredients that finally brought it out so explosively.

Furthermore, the debate would have broadened then and there. Under those circumstances it might have received valuable contributions far earlier than was actually the case from experts in an obviously pertinent discipline—those who are specialists in infectious diseases and the control of dangerous pathogens.

For those who continue to believe that recombinant DNA technology is entirely the work of the devil, it should make no great difference how the news first got out, provided that it did so. For those who consider the work legitimate and worthy of being continued, it probably does make a substantial difference.

All of this diagnosis by hindsight can be no more than speculation, but there are not many ways in which the word could have reached the general public that would have been more incendiary than the actual events.

The situation would have been worse, of course, if scientists had tried to hush up the controversy and keep it from reaching the public at all. This point will seem perfectly obvious to most practitioners of the press. I hope it is as obvious to everyone else. The scientists who contributed to making the issue public in the first place did so from a praiseworthy and extraordinary sense of public responsibility. Some of them today, one suspects, feel like a kindly person who volunteers to help an elderly lady across the street only to be mugged by the beneficiary on reaching the far sidewalk. The Catch 22 element in the situation was that these same scientists were probably doomed to reap the whirlwind sooner or later no matter what action they took or avoided. A strong public response may have been virtually inevitable simply because of the revolutionary nature of the research.

To summarize: an issue of this importance simply could not have been kept quiet indefinitely. As it turned out, representatives of three major newspapers knew about the letter before it was published and made it public in advance. Even if only one had done so, the subject would thereupon have become public and the circumstances would have been even more freighted with controversy and suspicion than was actually the case. In retrospect, it seems surprising that the subject remained internal to the scientific community as long as it did. By the summer of 1974, no reasonable person would have had anything to gain by keeping it internal any longer.

After the events of July 1974, the next milestone, of course, was the Asilomar conference. Many retrospective questions have been raised about that meeting, including the adequacy of arrangements for coverage by the press. Those arrangements have been criticized, among other reasons, for failure to make room for everyone who wanted to cover it. The following points can be made:

First, the concept of a pool arrangement, in which a small number of writers are delegated to act for the community, is reasonably respectable in the business and is not even particularly novel. It would have been preferable to open the occasion to any who wanted to come, provided space in the meeting rooms permitted. It also might not have been practical.

If one does opt for a pool arrangement, the choice of reporters for inclusion in the pool is important. Usually the list includes the two major wire services, AP and UPI, and at least a representative selection of the major newspapers and magazines. The selection should also include those writers who have been most diligent in covering the subject at hand, particularly those who have covered it from the beginning. My impression is that these fundamental rules were not entirely honored in the Asilomar arrangements.

In a world of news gathering in which the television networks are extremely important—as they are today—a separate pool arrangement should be made with them. In the case of Asilomar this may have been impractical for a reason that I will address later.

Before doing so, however, I point out that I have worked for newspapers and have written for magazines and radio, but not for television. I am therefore only familiar enough with the tribal customs of television to realize that they are different in some ways from print tribal customs. Radio is closer in many respects to the print media than it is to television, and it is for this reason, among others, that I use the term "the press." Nothing I can say about the dominant medium of the electronic variety is more than a layman's view.

In general, while pool arrangements are sometimes necessary, it is far simpler and more equitable to allow anyone to attend, providing space permits. Anyone not invited and not allowed to attend such a meeting is likely to complain.

The other feature of the meeting that has produced complaints, at least after the fact, was the ground rule that nothing be written until the meeting was over. Under the circumstances, that rule seems to have been justified and wise. Some of the writers who attended were not very experienced in covering molecular biology. It is a reasonable guess, bolstered by a little information, that many were pretty much ignorant of the subject and were getting their first introduction to it under extremely difficult circumstances. It was a complex subject, difficult enough simply to translate and digest for consumption by a lay audience. On top of that difficulty, the writers were exposed simultaneously to a controversy marked by sharp differences of opinion among experts. And, in debating those differences, the protagonists tended to dive into esoterica and jargon as soon as the debate grew hot. At least that is the way it must often have appeared to the writers. To have tried to cover this as a running news story on a daily basis would have been a recipe for disaster.

One suspects, furthermore, that the write-only-at-the-conclusion rule would be calculated to weed out at the start coverage by any newspaper or magazine that did not take the issue seriously. That can only have been a benefit.

One further reflection on Asilomar as a problem for the press: the meeting lasted several days and ended on Friday, with the release date for stories on the Sunday thereafter. To those who have not been faced with that kind of deadline, this may seem like a delightfully leisurely pace calculated to produce careful, reflective, and well-digested stories. In fact, for at least some reporters, it presents an agonizing time bind. For example, the copy deadline for the main news section of the Sunday *New York Times* is early afternoon. For Victor McElheny, covering it for the *Times,* the deadline was no later than mid-afternoon Eastern time; and there was a three-hour time difference between the clocks at Asilomar and the only clocks that matter to a *Times* reporter—those on the third floor newsroom of the *Times* at 229 West 43rd Street, Manhattan.

After Asilomar, the main scene of action moved to the National Institutes of Health in Bethesda, Maryland, and almost simultaneously became nationwide. By now, the main events are certainly well known to everyone. The following comments will concentrate on some of the aspects that most directly involved the press and some questions that come to mind retrospectively.

The first set of questions can be considered together: Did the press sensationalize and overplay the biohazard issue? Did the press have a tendency to focus on extreme views, and if so, did this contribute to the polarization?

First, a general comment. The press is not, and never has been, a single organization speaking with a uniform tone of voice or a unified point of view. The idea that the press is a monolithic single entity is only a delusion of zealots. Unfortunately, it does seem to be a common delusion, but no real understanding of the role of the press is possible in the presence of this misapprehension. Even the coverage of one issue by one newspaper is likely to vary from day to day and even from section to section of that newspaper. When the whole universe of news publications is considered, the variations can be tremendous. Nevertheless, it is possible to comment on the questions if one avoids the implication that the comments are either definitive or all inclusive.

On balance it seems to me that the press did not greatly overplay the biohazard issue. In fact, the biohazard issue in its broad sense was the only real issue. The press did not invent it, we were handed it in the letters to *Science* and the other journals, at Asilomar, and in many ways and forms thereafter. The whole purpose of the NIH guidelines, after all, was to safeguard against biohazards.

Indeed, there are some critics of the press who would assert that we underplayed the issue and have, somehow, conspired to do so. I do not agree. One could hardly have covered the controversy at all without giving prominence to the question of biohazards. On the other hand, the biohazards were, and still are, hypothetical and potential. It would have been irresponsible to paint them in any other light.

Beyond that, it must be remembered that the press seldom if ever makes issues up out of the whole cloth. We are, in various ways and various degrees, the pipeline through which issues and views concerning them get to the general public from many sources. If the press discusses biohazards, it is usually only by quoting the opinions of others. Some of these quotes will be responsible. Others may be irresponsible. In time some of them will prove to be false. But there are limits beyond which the press does not have the right to ignore views, even if their rational basis is suspect. Sometimes the seemingly irrational proves to be the truth.

It is true, as one of our most famous Supreme Court justices said, that freedom of speech does not permit one to shout "fire" capriciously in a crowded building. But it is also true that one has a right to stand outside a building and distribute leaflets charging that the building is a firetrap. That right holds, subject to libel laws, even if the charge proves to be false. There were distributors of such leaflets in the recombinant DNA controversy, and it was the duty of the press to cover their views as well as the views of others who had more comforting messages to deliver.

At this point, to illustrate the problem of sensationalizing from another angle, consider this quotation—out of context no doubt—from a paper given in 1977 by Dr. Charles A. Thomas, Jr., of the Scripps Clinic, then at Harvard. The title of the paper was "The Fanciful Future of Gene Transfer Experiments."[3]

Thomas quotes one specialist—not a journalist—as asserting: "Recent advances in biology and medicine suggest that we may be rapidly acquiring the power to modify and control the capacities and activities of men by direct intervention and manipulation of their bodies and minds. Certain means are already in use or at hand, others await the solution of relatively minor technical problems, while yet others, those offering perhaps the most precise kind of control, depend upon further basic research."

He quotes another as follows: "And it is true that the possibility of altering or replacing specific genes would significantly broaden our control over the human gene pool, compared with our present dependence on the lottery by which each of us forms an enormous variety of germ cells."

Of such remarks, Dr. Thomas says cheerfully: "We have all grown so accustomed to reading such statements from journalists who don't know any better that we now accept them from scientists who do know better."

At another point in his discourse, he says: "Our journalists (who are now more numerous than ever) prematurely publicize genuine scientific advances in terms that are consistent with those of science fiction. In this process, they often have the full complicity of the scientist himself. Who doesn't like attention? We should not be surprised to see scientists saying irresponsible things in order to enjoy public attention. It is human nature, and scientists are all too human."

In fact, there are many journalists who do know better and who are, believe it or not, conscientious, albeit also human. But there is another issue of integrity at stake when it is put in the context of the recombinant DNA controversy. Some of the witnesses at public hearings, such as that held by the National Institutes of Health early in 1977, put important points cogently. Others said things that seemed, to any critical mind, highly debatable and dubious at best. But the witnesses seemed to have made their points sincerely, at least, and presumably they were convinced that their views were vital to the great debate.

To what extent is a journalist justified in ignoring such views simply on the grounds that he or she does not believe them? The answer, of course, is that sometimes one can and sometimes one cannot, but the choice always has to be made carefully and conscientiously, bearing in mind George Bernard Shaw's old dictum (also taken out of context and, worse still, paraphrased): Do not necessarily do unto others as you would have them do unto you. Their tastes may not be the same.

Furthermore, it must be kept in mind that no journalist considers this kind of decision easy. It is all too possible to lead the reading public astray by overplaying the views of one or another faction in a controversy, and the cautious approach of playing it down the middle by quoting equally from both extremes is not necessarily a safeguard.

But that points to one of the prime dilemmas of covering an important and controversial issue such as recombinant DNA technology became in the middle and late 1970s. Fairness sometimes makes it necessary to quote some things that have become significant parts of the debate even if the writer suspects that some of those things are not true.

In the real world, if Mayor Alfred Vellucci of Cambridge, Massachusetts, expresses worries about bizarre things crawling out of a laboratory at Harvard, he is going to be quoted. Perhaps that does sensationalize the biohazard issue, but the press would be remiss in its obligations if it ignored him. But that statement puts the matter in much too fussy and academic a form. The press was simply not going to ignore him, obligation or no obligation. His remarks were news.

Just what constitutes news is almost impossible to define, but most experienced newspaper people have little difficulty in identifying it to their own satisfaction—and to that of their editors—when they actually encounter it.

And that raises another point often unappreciated by people outside the business. After a person, the mayor of Cambridge for example, has repeatedly made comments on a given issue, these comments usually become less and less compelling as news. No one in the business feels the obligation to continue quoting any spokesman for any issue forever if the comments remain the same or the situation evolves to make those comments less significant than they were originally. I will return to that point subsequently.

Meanwhile, we in the press quoted spokesmen for all manner of views on recombinant DNA if they seemed at all pertinent to the issues. While doing so, most of us were well aware that the various warring factions were trying to use us for their own advantage. They always do.

That is all part of the normal situation. To some extent the responses by the press probably did contribute to sensationalizing the issues. One of the practical realities of press coverage of a controversy is that the extreme views on each side are likely to be covered, and being covered, will probably eclipse the middle view in the attention of many readers. Whether this is justified usually defies determination until after the facts have long been interred in microfilm. It is not necessarily true that the middle view is correct. Sometimes it proves to be blatantly wrong. The extreme views at least give the reader a feeling for the outer edges of the debate. The reader probably needs to know those outer edges as well as the mainstream.

This problem of giving perhaps too much attention to the extreme positions is by no means exclusive to the recombinant DNA controversy. It is a continuing and difficult problem for which there is no foolproof answer.

Furthermore, as the recombinant DNA debate grew hotter and less factual, most of the writers covering it were painfully aware that some of the events and confrontations that played a part in the debate and the polarization thereof were nothing but media events. These, by definition, are events stages only in the expectation that they will be covered by the news media—and totally meaningless unless they are covered. In my opinion, the writing press is somewhat less susceptible to this kind of attempted manipulation than is television. The shouting of slogans and waving of placards is much more impressive on the screen than is any description of it in print. For television such window dressing is often, one suspects, harder to avoid than it is for writers. We can relegate it to a paragraph or so if we think that is all it is worth. For the television viewer it may sometimes appear to be the whole story, and not necessarily through any fault of those responsible for the newscast.

For some readers and some viewers outside the scientific community, such things as those described above probably did contribute to a distorted view of the issues. I think most of us did out best to avoid this, but it is only possible to provide the facts, so far as they are available, and the range of issues. We have no way of compelling a reader to read beyond the headline or read a story with enough attention to appreciate what it is designed to convey. We have absolutely no way of making a reader set aside his or her prejudices in the reading, although these may color the perception of a story in strange and wonderful ways.

The recombinant DNA controversy seems to have been a particularly difficult problem for the news media because its foundation is a complex area of science that, in itself, is often taxing to explain to a lay audience and because the debate started with a great paucity of reliable facts. All manner of scary scenarios were put forward, sometimes with implications that their fruition might be right around the corner, or at least the corner after next. When dramatically opposed views come with approximately equal vehemence from spokesmen who are equally august winners of the Nobel Prize, the press is at a disadvantage.

The recombinant DNA affair was by no means the first in which this difficulty appeared, and it almost certainly will not be the last. In the long run, such confrontations have a way of sorting themselves out so that one can make a valid judgment as to what is probably factual and what is science fiction. This cannot be done all at once, and it is most difficult in the opening stages of any controversy. There is no pat answer to the problem for either the press or the scientific community. Dealing with it takes patience, persistence, and a lot of hard work.

One useful aid to sorting out the validity of the issues is that of following the statements of the protagonists over time. When point of view and exhortation remain constant while the supporting reasons, logic, and alleged facts change to fit new circumstances, one can only conclude that the person is arguing mainly from faith and conviction and that facts are not necessarily paramount.

To summarize this aspect of the matter, the press probably did contribute to distorted views of recombinant DNA technology on the part of some readers and viewers. That part of the press with which I am familiar did not do this deliberately, and indeed, took great pains to avoid it. The essence of the problem can be seen, perhaps, in the fact that even today some observers of the press think we are poisoning the public's mind against the field, while others are equally convinced that we are simply tools of the scientists, parroting an onwards and upwards line put out by some scientists for their own crass aggrandizement. On balance, it seems unlikely that the press played any significant role in creating the actual polarization of views into conflicting camps of those who favor and those who deplore the research and, more important, its applications. The polarization was already there, waiting to be tapped.

Some elements of the press have covered the progress of molecular biology for many years. Even before the events of 1973, 1974, and thereafter, there have been some controversies, and we have covered those at least when they erupted in public. Since 1974, however, the amount of coverage by the general press has increased markedly.

Out of curiosity over the dimensions of this, I ran a rough check of the number and kinds of stories on this topic in the *Times* files. The figures that follow are certainly not statistically significant or even more than marginally accurate. But they do contribute to an impression. Before 1974, most of the stories filed under the catch-all label DNA were on one or another aspect of the scientific work. Only a very few were on any controversy involving it.

Through 1974 and up to September 1977, the number of stories greatly increased and their type, so to speak, was changed. My rough tally was that the *Times,* during this period, published forty-six stories on the recombinant DNA controversy, twenty-one that were fundamentally on recombinant DNA science, two editorials, two book reviews, and two letters to the editor congratulating the *Times* on an article by Dr. Liebe Cavalieri in the newspaper's Sunday magazine and two denouncing the magazine for the same article.[4]

During the next easily definable time period, September 1977 through the end of August 1980, there were twenty-seven stories on the controversy, thirty-one on the science, five editorials, two book reviews, and a new category—eleven stories related to the plans, prospects, and acts of industry.

For the period of approximately another year after September 1, 1980, there were seven stories primarily on the controversy, nineteen on the science, one editorial, and six stories on the industry. Since then the number of industry, and its reflections in the stock market, has increased substantially.

Of course, these figures do not prove anything. What they suggest, however, is a substantial waxing and waning over time in coverage of the controversy that has surrounded developments in recombinant DNA technology.

In the stories that are not devoted primarily to some aspect of the controversy, such as violation of NIH guidelines, there has also been considerable diminution of space devoted to the potential risks of the work. In the early days of the controversy, the division between potential benefits and potential risks was often about fifty-fifty. That seemed to be a prudent proportioning because these risks and benefits were all equally potential and, for the time being at least, equally hypothetical. That is no longer true. The potential ill effects are still potential and hypothetical, but the applications are becoming more and more a matter of reality. Under such circumstances it would hardly be rational to continue to give equal time to risks in stories that were not focused specifically on issues of risk.

It is also a reasonable conjecture, however, that the first tangible evidence of real risk or scientific impropriety would—will—be covered quickly and with vigor. Nor is that entirely conjecture. There have been two instances within the recent past of what seemed to be violations by scientists of the NIH guidelines. The press covered these very actively. In at least one of the cases, the issues may have received more attention than they deserved. In any case, that recent track record suggests that there will be vigorous coverage of any serious problems that arise in the future.

Recently some published criticisms of the press have also noted a waning of coverage on the controversy, so far as that controversy relates specifically to health effects. There have been charges that this trend in coverage reflects a situation in which the press is being lulled into complacency by the propaganda of the vested interests. It is also possible, however, that the more lurid aspects of the potential risks are no longer as credible as they once were.

There are fads in the attentions of the press, just as there are in most other aspects of human endeavor, including science. We have been through a fad in which the statements of science tended to be taken as gospel. We have been through another period characterized by a comparable uncritical acceptance of the statements of environmentalists and those who declared themselves advocates of the public. As always, these fads have not been uniformly followed by the press, but have been discernible in the total coverage of the news.

For the future, it is clear that the science press needs to be careful to avoid faddism. The history of the recombinant DNA controversy makes this clear. We must always be alert to the danger of being taken into camp—either by the persuasive arguments of the proponents of a field of work or by what one might call the "Ad Hoc Coalition for the Support of Chicken Little."

One of the questions likely to arise in such a controversy is whether the press views scientists as a special interest group or as a public interest group. The answer, of course, is both. Some elements of the press will take one view; others the opposite. Furthermore, it is quite clear to most of us that a scientist can represent a special interest on one issue and the public interest on another. For that matter, some groups that exhort under the banner of public interest are really representatives of one or another special interest. Labels make very little difference. It is performance that counts.

Has the recombinant DNA controversy been a blessing or a curse for science in particular and society in general? This too is a question on which a broad spectrum of answers is possible. In the final analysis, however, the question is irrelevant.

The recombinant DNA controversy has been an inevitable consequence of the rapidly increasing ability of science to study and manipulate the genetic material of our species and all others on earth. This has frightened some people, raised hopes in others. In general, it seems to have raised, at least for now, the level of public consciousness concerning important matters that used to be so far removed from the common experience and understanding that many lay publications simply ignored them and many lay people maintained a comfortable ostrich-like oblivion. So far as recombinant DNA work and molecular biology in general are concerned, those attitudes are no longer possible. The number of Americans totally ignorant of the principles that govern the chemistry of their own genes is smaller than it was before the DNA debate exploded. Citizens are not likely to go back to that general state of pre-Asilomar innocence.

Presumably, scientists have learned some valuable lessons from the controversy. Certainly they have learned, from the Cambridge, Massachusetts, experience of a few years ago, that lay people can learn about such esoteric matters as recombinant DNA and molecular biology and that they are capable of making rational judgments on the basis of the facts.

So far as molecular biology is concerned, it is unlikely that any controversy will erupt again with the same force and confusion that characterized the debates of the late 1970s. Controversy may continue—over such things as attempts at gene therapy, for example, and the increasing industrial applications of genetic manipulations. There is also an important issue in the relationships between industry and academic science that are growing out of this field. But the degree of ignorance and the credulity that characterized some of the public attitudes of a few years ago are not likely to return to this particular arena of human debate.

One hopes that the scientific community has also learned that rationality has a better than average chance of emerging victorious in the long run, if the press is given access to all sides and ramifications of a controversy. But there would seem to be one clear danger for the future: that the scientists who took the extraordinarily responsible step of voicing their concerns will have judged that they reaped not public understanding and appreciation, but the whirlwind. There could, accordingly, be a natural tendency to vow never to do that again, but to keep the controversies of science safely sealed off in the world of science itself.

One hopes, above all, that such an attitude will never prevail. The experience of the past several years should have taught all participants in the controversy that it is neither possible nor desirable to keep controversial matters hidden from the public. The public does have a right to know—and to worry—about such issues. The better the public is informed and educated, the more rational their response is likely to be. It could be argued that one of the seeds of hysteria over recombinant DNA work was widespread public ignorance concerning molecular biology and the chemistry of genetics.

Over and above the fact that the public has the obligation to be informed and the right to know, no issue as important as this one will ever in the long run stay hidden. Neither public understanding nor rational public response will ever be well served by attempts to keep the public in the dark, no matter how well intentioned those attempts may be.

Notes

1. Maxine Singer and Dieter Söll, "Guidelines for DNA Hybrid Molecules," *Science* 181, (21 September 1973): 1114.
2. Paul Berg et al., "Potential Biohazards of Recombinant DNA Molecules," *Science* 185, (26 July 1974): 303.
3. I have deliberately left unidentified the persons Dr. Thomas quotes because my use of them is, perforce, doubly out of context. For those who wish to pursue the matter further, the paper was presented in *Genetic Interaction and Gene Transfer,* Brookhaven Symposia in Biology, no. 29 (1977).
4. The magazine article was an account of postulated future horrors by Dr. Liebe F. Cavalieri of Memorial Sloan-Kettering Cancer Center. A substantial number of molecular biologists objected to it, while opponents of recombinant DNA work quoted the article widely. Sometimes it seemed that they were also trying to bolster the legitimacy of their position by linking the newspaper's name to Dr. Cavalieri's views.

The influence of the press extended to all quarters in the controversy. Particularly important were the perceptions from within the scientific community. There the widespread view was that the press had not fairly presented the true nature of genetic manipulation and had exaggerated the risks and implications, thus needlessly alarming the public. But it is also true that some of the writers of the doom scenarios had scientific credentials, so public education was not a simple matter.

Norton Zinder, a professor at Rockefeller University, was a member of the original Berg Committee on Recombinant DNA Molecules, Assembly of Life Sciences of the National Research Council, National Academy of Sciences, which recommended policy for the conduct of these powerful new techniques. He remained active throughout the controversy, fighting against what he saw as needless restrictions, and trying to influence public policy at the state and national levels. His candid account captures much of the color and intensity of what he was fond of calling the "recombinant DNA war" when it raged out of control.

7

A Personal View of the Media's Role in the Recombinant DNA War

NORTON D. ZINDER

Harold M. Schmeck, Jr., has aptly described the trials and tribulations of a reporter trying to cover a complex story. He is a professional and views events from that point of view—on high. However, my memories of the media's role in the recombinant DNA story are of the consequences it produced and the tone that it set while disseminating the news. It is against a background of these events and the concerns and actions they generated that I describe the role of the media. Let me say first that, with a few notable exceptions, the texts of the news stories—and the editorials in the major newspapers and magazines—were as balanced as one might expect for so complicated and volatile an issue. It was in other areas that the media often forgot its power and thereby became part of the problem.

The coverage of this story can be divided into three periods: Asilomar (1974–76), the "recombinant DNA war" (1976–78), and detente. The Berg letter (1974) alerted the press to the existence of molecular biology and its newest progeny, recombinant DNA. Other than noting that over the years most of the Nobel Prize winners in medicine were molecular biologists, the science press was too taken up with space, cancer, and viruses to note this growing discipline. Despite the excitement of the headlines generated by the Berg letter—along the line of "Geneticists Ask for Ban—World Peril Seen"—the press left the public with just a hint of important events, but without explanation. Too few of the reporters understood science well enough to be able to examine all of the stories' implications.

The press coverage at the Asilomar conference (1975) was unique. It was pooled coverage with a proscription on filing any story until the meeting was over. This had significant consequences. First, the stories were excellent. Reading them today I can almost relive the experience. Reporters, who knew little of science and stayed four days with the scientists listening to the arguments, learned not only about the science but also about the politics of science and the scientists themselves. Their stories reflected their encounter with a sincere, hard-working group struggling with an almost insoluble problem, that is, how to protect the public in the most responsible manner when it wasn't clear if there was any hazard to protect against. A secondary consequence of this mode of coverage was that the reporters got to know us personally and came to respect our sincerity if not our wisdom. When, later, the atmosphere changed and recombinant DNA research was under attack, they did not believe we were the self-serving, venal scientists that some made us out to be. The tone of all future press coverage was modulated by this shared experience.

After Asilomar, the tempo of events quietened, as did the coverage, with only an occasional in-depth piece appearing in a magazine. Only Nicholas Wade in *Science* magazine consistently followed the ups and downs of the writing of the NIH guidelines.

In the winter of 1975–1976, there was a challenge to recombinant DNA research at the University of Michigan. Hearings were held, and the work was then allowed to proceed. This occurred without notice other than in the school newspaper, although material had been made available to the wire service by the University Information Service.

What does it take to make a media event? In June 1976, the NIH guidelines were promulgated but were scarecely noted because of the beginning of the recombinant DNA war in Cambridge, Massachusetts. Major Alfred Vellucci of Cambridge, aided by an assortment of biologists and Boston ideologues from Science for the People, held hearings on the safety of recombinant DNA research. The event had been generated by the desire of Harvard's biochemists to build a P-3 facility. The biologists and biochemists at Harvard now had a cause celèbre with which to escalate their long-time mutual antipathy. P-3 and P-4 research was banned in Cambridge until it could be reviewed by a committee of "public safety." The press responded with alacrity as Vellucci and others made outrageous charge after outrageous statement and all in colorful language. This was news at its best—town against gown, with some gown in favor of town. The hearings even drew the attention of television: eminent scientist against eminent scientist in a hearing chaired by a old-fashioned, gruff-speaking politician. The issue almost paled before the drama: Just say something threatening about DNA research and Vellucci grunts approval. The refutation is technically too difficult for most people to understand, and hence the idea of danger is fixed.

The detailed coverage of the Cambridge hearing led immediately to concern in other localities. The day the *New York Times* published the story, I was taken aback when a shrill voice shouted at me over the phone, "Who's doing those experiments in this state?" Thus began a two-year battle with the New York State Attorney General's Office and, ultimately, the New York State Legislature over legislation regulating recombinant DNA research.

Shortly after the hearing in Cambridge, a second media event exploded. Liebe Cavalieri's article, "New Strains of Life—or Death," appeared in the *New York Times Magazine*.[1] Recombinant DNA research was touted on the magazine's cover as a potential holocaust, and inside was an article which set forth many pages of horrors. Cancer would be set rampant in the population by self-serving, elitist, Nobel prize-seeking scientists. The stature of the *New York Times* and the fact that the article was written by a scientist from the Memorial Sloan-Kettering Cancer Institute both gave the article credibility. The author, a self-avowed antielitist, was taking advantage of his elitist position and institution to confound the elite *New York Times*. There were reverberations all over America. A month later, the article was introduced into the *Congressional Record* by Senator Jacob Javits. Although this garbage was in the media it was not of the media. It took a few days for me to contact Victor McElheny, who had been moved, in the Machiavellian ways of the *New York Times*, from covering science at the city desk to science at the financial desk. He was no longer on the story. This was tragic, for of all the reporters, he was not only the most knowledgeable but also the first and most vigorous in covering this area. Vic explained that the *Times Magazine* was an empire unto itself, that none of the science writers of the *Times* had seen the article, and that they were all as dismayed as the rest of us.

From that moment on, the situation began to go downhill for us. Who is us? We are the authors of the Berg letter, the organizers of Asilomar, and the vast bulk of the molecular geneticists. The articles in the media now reflected the strong opposition to this research by a small number of dissident scientists and certain public interest groups, particularly the environmentalists. The fall and winter of 1976 saw more and more articles in the newspapers and magazines attempting to discuss the pros and cons of research with recombinant DNA; a few even touched on its philosophical implications. The texts themselves were not unreasonable; they tried to be balanced, but charges of dire consequences make better copy than the more technical counterarguments. It is a well-known political fact that once a charge is made it is almost impossible to expunge. One of the great ironies was that those of us who had called the issue to public attention were now being portrayed by the press and others as the adversaries of the opponents of this research. However, most of us had not really changed our position, which was to move forward with due caution as determined by new results. Due caution, however, had now become a negotiable state. Clearly, to some we were either stupid or venal, as they proclaimed our carefully wrought guidelines for performing this research inadequate if not useless. They implied that to trust us to regulate ourselves would be positively unAmerican. this came about in part because several environmental organizations had decided we were the enemy. Their political know-how, national audience, and use of the adversary procedure further publicized and intensified the debate.

Despite the continued relatively balanced texts, the tone and the mood emanating from the media changed drastically. The headline writers, cartoonists, and illustrators had a field day. Creepy, crawly creatures abounded. A sober article in the *New York Times* by Stephen Toulmin, a professor of philosophy at the University of Chicago, on the constitutionality of regulating DNA research ran beside a drawing of a test tube filled with crawlies running the full length of the Op-Ed page. A DNA helix was topped with a snake's head dripping venom in the *New Times*, while a balanced article in *Time* magazine was illustrated with a scene from the movie "The Demon Seed" showing Julie Christie being raped by a computer. Cartoonists drew bug-eyed monsters about to attack someone; the sign behind said "Genetic Research." The general public, which undoubtedly found many of the well-written technical and philosophical arguments hard to follow, could not miss the implications of these accessory elements. That something scientifically important and, despite all, perhaps beautiful had happened was lost in the transmission.

During the next two years, my colleagues and I were much in demand to speak to the public at meetings in churches, synagogues, universities, and other places. The audiences were not hostile, just uncertain; a bit insecure and wanting to be reassured. The questions led us to believe that with careful explanation, people became more interested than frightened.

When congressional regulatory legislation was pending (1977–79), the reportage was essentially straightforward. The *Washington Post* tried to be helpful by suggesting in an editorial (4/12/77) that Congress listen to my words of caution before moving into the regulation of scientific research. But the media only intermittently followed the ebb and flow of arguments for uniform national regulation pushed by the proponents of recombinant DNA research, versus those for local option, pushed by the opponents. Almost all scientists agreed upon national, uniform regulation, except some ideologues and environmentalists. A hodge-podge of local laws could have destroyed the science. The intensity of our feeling on this issue was never transmitted by the press, but neither was the similar intransigence of those in opposition. Moreover, it was this intransigence that provided the scientists with the lobbying points they used to kill all of the pending legislation. Remarkably, to my knowledge, no piece analyzing the failure of the legislative effort was written. (See Chapter 3 for additional discussion of this subject.) Because of its congressional locus it was the central event of the dispute.

In the summer of 1977, Walter Sullivan, the dean of *New York Times* science writers, wrote an analytical piece on the status of current opinions on recombinant DNA research. He described new scientific data; resolutions from the Gordon Conferences against legislation; a letter from Roy Curtiss, of the Microbiology Department of the University of Alabama Medical Center in Birmingham, recanting his previously negative position; and the conclusions from the Falmouth Conference on Risks of Recombinant DNA. It was a thoughtful piece, and Sullivan concluded that although many questions remained, the consensus among scientists was that the research was safe enough to proceed. I was particularly pleased because I had supplied him with some of the documents and a few moments of discussion. When Sullivan's article appeared, the opponents angrily maintained that many of these documents had been exaggerated or were premature. However, these opponents had lost most of their constitutency among scientists, and as they repeated the same tired arguments, the press began to lose interest in them. The press always requires something new for its interest to be maintained.

Soon I was desperate for lack of access to the media. Harold Schmeck says that "various warring factions were trying to use us [the press] for their own advantage." I tried twice and failed. The first time was mid-summer 1977. In Washington, the Kennedy bill, with its monstrous bureaucratic commission and no preemption of local law, was moving to the Senate floor; in Albany, the governor of New York had waiting for his signature a very restrictive and highly ambiguous piece of legislation; the New York City Council had developed a resolution banning P3 experiments that was headed for hearings; and the Manhattan Borough Board was about to hold hearings on banning recombinant DNA research. The local resolutions had been eased through during the summer absences of councilmen, academics, and others. I wanted Schmeck to write the story of these machinations, but more important, I hoped that the public exposure of this story would in itself have a mitigating effect. I thought no matter what he might write, it couldn't fail to help. Moreover, Senator Kennedy might have noted the consequences of unrestricted local option from such an article. Also, it was in every sense a valid story that deserved telling regardless of one's position on recombinant DNA. Unfortunately, Schmeck could not do it. Fortunately, however, the Manhattan board hearings were recessed and never rescheduled, the city council never moved forward, and the governor vetoed the bill. How easy it is to write that sentence now—but at the time it was another story.

The governor's veto was terribly important. At that time, it was not clear whether the legal regulations, if any, would be local or national. Also, the political dynamite assumed to be in this issue for politicians who opposed the will of the opponents of the research was thereby defused. The *New York Times* noted the veto as a small item in the middle of its second section, where state legislation is always handled. At least it made the headlines of notable state events.

The second time I tried to impose on Harold Schmeck was a year later, after the legislative efforts had died away and the NIH had screwed up its courage to try to revise the guidelines. The opposition now was primarily the Friends of the Earth (FOE), the Environmental Defense Fund (EDF), and the National Resources Defense Council (NRDC), all with considerable political clout. We attempted rapproachement with them. However, Lewis Thomas of Memorial Sloan-Kettering Cancer Center in New York resigned from the FOE board in anger. Paul Ehrlich tried to arrange a meeting between Paul Berg and Dave Brower (FOE) and failed. James Watson tried to talk with EDF and failed. Discovering, oh so late, that my colleagues René Dubos and Joshua Lederberg were on the NRDC board, I met with the NRDC, but not before Dubos had resigned from the board and Lederberg had publicly disavowed its stand on recombinant DNA. When the staffs of some of these organizations were asked for the dates and locations of their board meetings in order that a discussion might be arranged with some scientists, the answer was that it was none of our business. They were private organizations, we were told, and we could not attend their closed meetings—this from organizations that call forth a cascade of paper via the Freedom of Information Act at the drop of a twig and insist that all advisory committees have open meetings. It was a story that should be told, not only to set the record straight but to provide a broader picture of the democratic process in these so-called public interest groups. (See Chapter 9 for a discussion of the role of public interest groups.) Since it was not a local story, when Schmeck did not do it, I called Barbara Culliton of *Science*, and shortly thereafter, Eliot Marshall wrote a fine article documenting our impossible relationship with the environmentalists.[2] It created a stir in many circles, and although only the NRDC dropped the issue, the others backed off.

There is a class of newspaper in this country, epitomized by the *National Enquirer*, which thrives on sensationalism. Recombinant DNA chimeras were made to order for them. The monsters formed and the horrors caused were limited only by the imagination of the writers. The impact of this form of journalism is hard to know. But we had no recourse other than to live with it.

The other media seemed to play small roles. I was asked to do four interviews and one debate on local radio. By clever splicing and a voice-over, one radio interviewer had me mouthing the opposite of my own position. I reviewed two scripts on recombinant DNA for public television that were atrocious; happily they were never produced. The most diligent coverage of the recombinant DNA issue was to be found in *Science* and *Nature*. Colin

Norman at *Nature* kept the story relatively straight, but when he wrote for *Science and Government Report*, a rather different point of view was reflected in the stories. Once he interviewed me as the correspondent from *Nature* and produced a positive article there, but in *Science and Government Report* I found the same information scarcely recognizable.

Nicholas Wade of *Science* was not only at Asilomar, in fact he had covered the story from its original with the Singer-Söll letter. As the story progressed, it became subtly twisted in an anti-recombinant DNA fashion. Although we did not expect *Science* to be a house organ or Wade to be our shill, we did expect a fair shake in *Science*, if nowhere else. His influence was wide, as smaller newspapers without real science writers used the articles in *Science* for guidance. Even in the People's Republic of China, as I discovered in the middle of these events, the result of his bias was apparent. The only English-language journal the Chinese read was *Science*; they knew the technical details from the excellent special issues of *Science* on recombinant DNA research, but were totally confused by the politics of gene splicing.

Wade had reported favorably on the formulations of Robert L. Sinsheimer, then chairman of the biology division at Caltech, about scientists taking over evolution and transgressing the postulated natural barrier to genetic exchange between eukaryotes and prokaryotes. No one would take on Sinsheimer on his own terms, Wade proclaimed. This is not the place to disprove the hyperbole. However, after Wade had interviewed Sinsheimer, he talked with Maxine Singer of NIH and myself on the precise issues he later wrote about.[3] Our replies may have seemed too inadequate to quote, but they did exist. Just one other example of his bias: Wade wrote, "The strength of reaction against the second wave of critics seems in some ways disproportionate. . . . Most of their arguments in favor of stricter guidelines, whether right or wrong, are not inherently extreme or unreasonable."[4] To my mind, anyone who could write that in 1977 lived in another world. Even a casual glance at the NIH volumes on recombinant DNA research[5] reveals how unreasonable a statement that is (see also M. Rogers[6]). Moreover, the statements in these volumes had been prepared for polite debate as letters and recommendations to the director of NIH. "We propose the NIH withhold all funding until the level of safety had been established." "I would propose that *all* recombinant DNA work be performed under the highest possible precaution at *one site* in this country (Fort Detrick)." Of course, it does depend on what one means by the word reasonable. To some, stopping *all* recombinant DNA research by making it impossible to do experiments seemed most reasonable.

The Sinsheimer formulations, particularly the postulation of a boundary between species, captured the public's imagination. Most people didn't know what the boundary meant, but felt that violating it must be like invading Canada. Over and over the press repeated this mysticism. This evolutionary argument was difficult to counter, for, as Sinsheimer phrased it, "How can we know what has happened behind these boundaries for millions of years?" This is unanswerable and is only one example of the question starting "How can you be sure . . .?" which was always asked.

The dissident scientists continuously asked what Professor Carl Cohen has called "heavy" questions.[7] Since they came from credible scientists, the press dutifully reported them. These questions are not designed to be answered; they are loaded with assumptions and allegations and are highly compounded. Erwin Chargaff's "Have we the right to counteract, irreversibly, the evolutionary wisdom of millions of years in order to satisfy the ambitions and curiosity of a few scientists?" is not a real question; it is a set of tonal allegations. Professor Cohen describes dozens of these questions and brilliantly dissects them. However, they did their damage by sounding profound and ominous in their continuous repetition by the press. All attempts at answers were necessarily incomplete, too technical, and could not undo the false associations that had been made. When coupled with threatening headlines, dire illustrations, and cartoons, these words helped lead to a confused if not somewhat fearful public. The reporters were also taken in by these high-sounding words until after a time when their repetition became tedious.

If the battle turned bitter, it was because of our frustration with these heavy questions and the often gratuitous attempts of the opponents explicitly to scare the public. In the middle of a debate on New York radio, George Wald suddenly asked me "What have there been so many strange deaths in this country lately?" One can't debate with people like that! There was no way for me to recall, by any simple statement, the allegation that had just gone out to thousands of New Yorkers. He didn't mention recombinant DNA but at that time Legionnaire's disease was about and not yet understood. He knew what he was doing; he'd been fingering an article on this disease all through the debate.

Erwin Chargaff was one who sought immunity from accusations of ad hominem argument in the metaphor and the plural. "Although I do not think that a terrorist organization [us] ever asked the F.B.I. [the NIH] to establish Guidelines on proper conduct of bombing experiments [recombinant DNA] . . .".[8] When Jim Watson, who was notorious for speaking his mind, countered with "kooks, shits and incompetents" (as *Time* reported[9]), the opponents yelled foul. I can't imagine why. "Kooks" is as plural as "terrorist organization" and "shits" is as metaphorical as "bombing experiments" although not as felicitous. In characteristic fashion, the press never noted the many similarly disguised attacks on the personal integrities of the Asilomar scientists; but Watson's remarks did not go unnoticed.

Still another problem was posed by the media. Some writers used a special set of rhetorical devices to embellish their stories. There were analogies to Dr. Frankenstein, of playing God, and of course, the crossing of species boundaries. All these are somewhat amusing now, in that they show little understanding of evolutionary theory and are in fact creationist in implication, but they had their impact. They implied we were doing something occult; that recombinant DNA experiments were outside natural law. But there were and are no miracles involved, only simple principles of bacteriology, cell biology, DNA chemistry, and enzymology. Did they really think so little of God, or were they thinking of the devil? It always made me uncomfortable to be asked if we were playing God. The implication was that we could.

Overall, the media was probably as balanced as could be expected, and things are peaceful now. The media did try to follow and not lead the story. For a while, they had some trouble in deciding who were the good guys and who were the bad guys among the credible scientists. Since it is easier to attack than defend, they got trapped for a period in disseminating outrageous statements—which are newsier. It took a while to sort out the fact that when scientists take sides, they are not necessarily distributed normally. In this dispute, the ratio at first was about 80 for to 20 against, and shortly thereafter, 95 in favor of rDNA research to only 5 against. Minority positions need presentation, but the relative size of the minority should be made clear, especially on a technical issue. The media's major failure lay in their inability to explain to the public that recombinant DNA is just shorthand for the technique of splicing together DNA molecules. Techniques are neutral—something like driving a car. At issue was the kinds of molecules spliced together and the uses to which they were to be put. Thus, the thousands of different kinds of recombinant DNA experiments were all put in one pejorative bag. Had the media made clear that the majority of the experiments were agreed by all to be safe, we might sooner have gotten to the point of negotiating price, not mythic principle.

Prescient though it may be, the *New York Times* was five years premature in transferring McElheny to the financial desk. Today, much of the reportage on recombinant DNA and the biotechnology industry it spawned lies there. Also, a surge of public interest in science, possibly generated by the recombinant DNA debate, has led to the appearance of many new science magazines. Stories on the triumphs of modern molecular genetics abound. Harold Schmeck's columns now regularly contain phrases such as, "the hazards once thought to be associated with this research are now considered to have been exaggerated." Thus does the press write finis to the recombinant DNA controversy.

Notes

1. Liebe F. Cavalieri, "New Strains of Life—or Death," *New York Times Magazine*, 22 August 1976.
2. Eliot Marshall, "Environmental Groups Lose Friends in Effort to Control DNA Research," *Science* 202 (22 December 1978): 1265.

3. Nicholas Wade, ''Recombinant DNA: A Critic Questions the Right to Free Inquiry,'' *Science* 194 (15 October 1976): 303.
4. Nicholas Wade, ''Gene Splicing: Critics of Research Get More Brickbats than Bouquets,'' *Science* 195 (4 February 1977): 466.
5. ''Recombinant DNA Research: Documents Relating to 'NIH Guidelines for Research Involving Recombinant DNA Molecules.''' U.S. Department of Health, Education, and Welfare, August 1976.
6. Michael Rogers, *Biohazard* (New York: Knopf, 1977).
7. Carl Cohen, ''Restriction of Research with Recombinant DNA: The Dangers of Inquiry and the Burden of Proof,'' *Southern California Law Review* (1978) 51(6): 1081.
8. Erwin Chargaff, ''On the Dangers of Genetic Meddling,'' letter to the editor, *Science* 192 (4 June 1976): 938.
9. ''Tinkering with Life,'' *Time*, 18 April 1977, pp. 32–45.

A continuous theme throughout the policy discussions over the use of recombinant DNA techniques was the matter of who should make the decisions. It began within the community of molecular biologists, but quickly erupted as a public issue involving governments at all levels and university campuses. The University of Michigan was one of the first where the conduct of gene splicing on campus was debated. Donald Michael, a social psychologist now living in San Francisco, was at the time a member of the Michigan faculty and a prominent participant in the controversy.

Is it an issue to be resolved purely among scientists because it is technically complex? Is it a public issue, because untoward consequences might affect anyone, and therefore to be dealt with through the democratic institutions that govern our communities, states, and nation? Or should it be left to experts—those whose particular talents and experience make them uniquely qualified to make decisions of profound implications to the future of science and perhaps of mankind? Don Michael asks what is perhaps the most important question of all, Who decides who decides?

8

Who Decides Who Decides? Some Dilemmas and Other Hopes

DONALD N. MICHAEL

The Question in Its Context

After careful consideration, one part of a community seeks to undertake a scientific activity, in which it has deeply vested interests, that will put another part of the community at a problematic risk for what are believed by the activity's proponents to be socially worthy reasons. Who, and on what basis, decides whether the action is permissible? And who decides who decides when circumstances are sufficiently unconventional to raise questions about the procedures and legitimacy of conventional decision-making structures and personalities? Indeed, who decides that circumstances are sufficiently unconventional to merit new decision-making procedures and participants? What seems like an infinite regress results from the dissolution of accepted values and norms and accepted processes for establishing and maintaining them. This dissolution results partly from the influence of scientists themselves, some of whose words and works have helped define and extend the conflicting issues burdening this society and the world. Because of the pervasive yet ambivalent roles of science and technology, an important consequence of this dissolution of shared values, and of the decision-

Note: "Who Decides Who Decides?" was originally published in Stephen Stich and David Jackson, eds., *The Recombinant DNA Debate*, Prentice-Hall, New York, 1979. This version is revised and updated.

making procedures that represent and reinforce them, is an increased focus on the conduct and consequences of scientific research.

In this chapter I shall use the question of who should decide whether to undertake recombinant DNA research in a publicly supported university to illuminate the more general question: What persons and procedures should determine whether to undertake publicly supported esoteric science that is potentially hazardous? The recombinant DNA issue presages things to come in biology and possibly in the social sciences. The University of Michigan, a publicly supported institution, is a prime example of organizations whose existence depends on funds produced through taxes and which, therefore, must serve the public interest. To understand better the nature of the problem, I shall try to relate its abstract aspects to a real-life example, alternating between abstract exploration of the problem and attention to the University of Michigan experience.[1]

A conventional response to the questions raised earlier would be the following: any decision, being based on esoteric knowledge and intentions and being undertaken at least in part for the public good, should be decided by the scientists and administrators involved, probably with some opportunities for comments and suggests from the community at large. But in the end, the decision should be made by the conventional decision-making structure, which, presumably, has the best interests of all parties at heart. This is especially so with scientific research, because disinterested good will can be expected to prevail and new knowledge can be expected to advance human welfare. Furthermore, all that could reasonably be done would be to minimize the risks, but risk is part of life and part of the cost of gaining new knowledge from which humankind ultimately will benefit.

An alternative response, the one that undergirds issues explored in this chapter and one apparently subscribed to by growing numbers of interested citizens, argues that whatever level of risk is involved, those who might become victims must have a formal part in deciding whether and under what circumstances to accept the risks.

This argument continues: given the nature of recombinant DNA research, all people everywhere should have a say since adverse consequences of the research could well be world-wide in impact. At a minimum, the community immediately adjacent to the research setting should be directly involved in deciding whether the research should be undertaken, especially when the research would be supported by public funds and conducted in a publicly funded organization.

To appreciate better the argument for this position and the dilemmas and difficulties that arise when attempting to transform generalities into operational terms, it will be helpful to examine the risks associated with recombinant DNA research.

Two types of risks have been delineated: process risks and product risks. Process risks pertain to the consequences of *accidental dissemination of research substances* into the environment outside the laboratory. Product risks pertain to the consequences of *deliberate dissemination* into the larger environment of the *products* finally produced from the research effort. In the case of product risks, evaluation is based on whether the hoped-for but undemonstrated benefits of such research will outweigh the feared and unknown costs (where costs and benefits refer to more than monetary considerations). For recombinant DNA, the costs are unknown because of our ignorance concerning the interactions of these new, chimeric life forms with natural life forms and because this research has at least the potential for irrerversibly changing human life itself.[2] This enormously complex issue involves questions far beyond those associated with the costs and benefits of other technologies, but those questions are not the topic of this chapter—except to observe that beliefs about an acceptable long-run balance of product costs and benefits probably influence feelings about the degree to which process risks should be accepted in the short run.

Arguments about process risks have to do with how perfectly the laboratories and their biological contents can be insulated from the surrounding community and whether, if such substances leaked into the larger environment, they could be expected to have deleterious impacts. The esoteric issues of biology and of probability calculations involved in such assessments are not the topic of this chapter either, except to observe that it is generally conceded that (1) extant probability calculations assume ideal performance by all researchers and others in the laboratories and that such perfection cannot be expected, given human fallibility and the likelihood that emotionally disturbed persons will sooner or later become involved in these activities; (2) we really cannot estimate the consequences of accidental leakage of recombinant DNA organisms into the environment because we don't know what organisms would be leaked nor do we know enough about the environment to estimate what would be involved in coping with a leak;[3] and (3) if one disregards human fallibility, the odds of accidental dissemination from a specific installation are very, very small, although history amply shows that rare accidents do, in fact, occur.[4] So there are legitimate questions about just how small the odds would be in real life, and there are serious unanswered questions about the consequences of those low probability events if they should occur. The consequences could in fact be enormous and a major concern for both scientists and nonscientists. So for all these reasons, some in the unversity and in the larger Ann Arbor community saw a compelling need to face the question of whether or not to undertake the research using recombinant DNA techniques.

Given the problematic nature of the risks associated with esoteric research, very difficult operational issues will attend any new procedures that include the larger community in decision making about undertaking such research. There are sources of discontent with and even repudiation of decision-making procedures based on expertise and duly constituted authority, and of the overriding priority conventionally assigned to freedom of inquiry. The four sources of discontent I will examine are not the only ones, but they illustrate especially well the extraordinarily complex decision-making tasks with which scientific endeavors such as recombinant DNA research burden our changing society.

First, the ideological and psychological virtues of direct or at least less indirect citizen participation in decision making enjoy growing acceptance. The ideological argument asserts that participation is a right of any person or group that might suffer the consequences of unilateral decisions made by a formal organization. The psychological argument asserts that decisions can be improved and a sense of community enhanced if recipient publics participate in the formulation of the questions and the design of the answers. In this way the various publics come to understand the tasks and problems involved in making and implementing a decision and experience a deeper sense of responsibility and commitment to it. And most importantly, participation provides the occasion for examining—or discovering—ethical issues. By itself participation does not resolve these issues. It does, however, proivde an opportunity for learning new ethical norms.

A second source of pressure for new decision-making practices is a growing challenge to the autonomy customarily accorded scientific research, especially supported by public funds. Significantly, the challenge comes from both lay opinion leaders and accomplished scientists. Increasingly, questions are asked about what research should the public pay for (i.e., what research contributes to the public weal) and under what circumstances are scientific research and its resultant technologies appropriate to the human condition. There seems to be a substantial antitechnology undercurrent that, while by no means exclusively correlated with citizen participation, may often be found in close ideological association.[5]

A third challenge to conventional decision making is widespread questioning of the legitimacy of existing organizations, that is, their entitlement to make decisions affecting those outside the organization and the validity of the processes by which they do so. Also questioned are conventional definitions of what constitutes competency to make such decisions. Throughout society there is now much distrust of large organizations. Since scientists are mainly associated with large organizations, there is growing distrust of the image of scientists that shows them to be motivated exclusively by a disinterested devotion to truth. This distrust is amplified by a growing recognition of the intense competition among scientists, and of the heavy dependency on public funding of scientific research and of the organizations for which scientists work. There is awareness that deeply vested mutual interests may be reflected in decisions—decisions that may not seem right to the public that pays the bills and sustains the risks.

The fourth factor, exacerbating all the others, is widening recognition that science is not ethically neutral. Thus decisions regarding science and technology cannot be made exclusively in terms of scientific and technical arguments, even though these are critical to the decisions.[6] Inevitably, scientific and technical information will be incomplete, especially in new areas such as recombinant DNA research. What is more, the information available results from earlier decisions about what merited most attention and what could be learned with available time and money. So the available facts and other data are also expressions of the value judgments—or biases—of those who collected the data and those who funded their collection. Such judgments necessarily go beyond purely logical, technical issues into realms of political feasibility, esthetic norms, rightness, and goodness.

These sources of discontent with conventional decision making reflect, of course, a widely shared distrust of representative government's ability and interest in responding wholeheartedly to community concerns that may be at odds with those of institutions such as universities, corporations, or big science. This is not to say that there is no role for established representative government. Rather, the relationship between that role and that of ad hoc community action has yet to be delineated. Therefore, the focus here is on exploring the nature of the emerging issues and approaches characterizing the community interest—self-help—side of the relationship. Questions about how the two aspects of governance interlock must be deferred until the attributes of the new side are clearer both in concept and in action.

The University Community Engagement

All of these considerations were part of the dialogue at the University of Michigan concerning recombinant DNA research. The most explicit and dramatic opportunity for university and Ann Arbor community engagement were the forums: a series of well-publicized, carefully organized lectures and discussions in the spring of 1976 open to both the university and Ann Arbor communities. Local and national experts were involved, examining biological, legal, ethical, and sociopolitical facets of the recombinant DNA issue. My informal canvassing of those involved indicated that the forums were variously seen as mere ritual, as building a new consciousness about the relationship between publicly supported research and the surrounding community, or as a laboratory for developing new means of university-community decision making. Only a minority argued that Ann Arbor citizens should have a formal part with regard to formal decision making about whether to undertake moderate-risk recombinant DNA research. The duly constituted authority position was the dominant one within the university; nevertheless, the university administration and the scientist proponents recognized the need for an informed community and that the university might benefit from community advice. The administration had established Committee B (in 1975) to deal with recombinant DNA research questions, and the administration and the Senate Assembly funded the forums and established the committee that designed them. Some recommendations in Committee B's report urged that citizens be members of the research monitoring committee and a proposed oversight committee. Suggestions for citizen membership on these follow-up committees—rather than giving them a part in deciding whether there should be research at all—was the university's not unconventional response to this unconventional problem. This expressed, surely, the usual reluctance of those in power to weaken their prerogatives. But there was another consideration that preoccupied many and will continue to do so as the decision challenge recurs: protection of the freedom of inquiry.

Most university faculty members and, indeed, most educated people everywhere in the Western world believe that freedom of inquiry must not be constrained, especially not by persons outside of the community of peers associated with the inquiry. This belief is, however, increasingly challenged both by some who well understand its importance for an open society and by others more preoccupied with other priorities.[7]

From this perspective, if the university were to forfeit its exclusive right to determine whether recombinant DNA research—or any other type of investigation—should be undertaken, it would very likely be establishing a precedent with regard to freedom of inquiry in other areas where members of the community could also argue that they were being put at physical or emotional risk by the research process or its possible products. What with the changing attitudes toward science, participation, and decision making, such a precedent would profoundly disrupt the elaborate and subtle mechanisms that motivate and guide science and systematic inquiry in general. Consequences could be as unpredictable and possibly as societally catastrophic as those feared from DNA research itself. However, some would argue—myself included—that the very fact of growing challenges to freedom of inquiry and to its maintenance through duly constituted authority, makes it all the more necessary to begin now to discover new ways to reconcile the demands for participation by those putatively at risk with demands for protection of freedom of inquiry. Both demands carry heavy costs as well as great benefits. Recognition of these and therefore of the need for a new overarching ethic endowed the recombinant DNA research decision-making issue with both symbolic potency and unique potential for initiating learning what such an ethic might be and how it might contribute to decision making about such activities. It will take time and much experience to learn what values and techniques work, and the hour is already late. What then are the questions in need of new answers in order to agree upon "who decides who decides?"

Deciding Who Decides

The first question we could ask is, What is the appropriate geographic and temporal scope from which to draw the decision makers? With chimeric biological materials, it is impossible to anticipate how widespread will be the consequences for natural life forms.[8] Therefore, the appropriate decision-maker pool would seem to be the whole world as well as future generations since everyone, especially future generations, may be the beneficiaries and/or casualties of this research. But there is no such decision-making capability; the initial examination undertaken by scientists during a self-imposed moratorium, is as close to world-scale participation as we've gotten.[9]

Lacking world scale, or even a world regional scale decision-making capability, we are thrown back on the nation as more appropriate than the immediate environs around the research laboratory for decisions that might have profound consequences over space and time. The NIH guideline deliberations were an exceptional and on the whole admirable experiment in this direction, even though these lacked sophisticated studies delineating the long-term social costs and benefits of the research, in part because we know too little to do very much in this direction. Moreover, the question of who would be entitled to participate in decisions about process risk exposure near where the research would be done was left unexamined. Instead, the main emphasis was on how to balance the need to minimize risks for those outside the laboratory against the risk that if the constraints were too stringent some scientists would disregard them.[10] But the fundamental flaw in the NIH approach was that it reinforced the usual mode of operation wherein geographically separate institutions compete for funds and for the prestige won through successful research. This mode inevitably puts a premium on getting there "first with the most," and it focuses concern at the local level about whether to incur the associated risks.[11] (Even though a local accident might result in world-wide consequences, the acceptability of the risk probably depends on one's perceived geographic proximity to the source.) At the same time, those seeking to do the research are acutely aware that they are in competition with scientists in other locales who may not be delayed by local demands for community involvement. Therefore, perceived localization of risk by some and pressures to get on with the research by others can be expected to be a likely setting in which new forms of decision making will be created and implemented. That context is assumed in what follows.

Under these circumstances, who then should be involved in decision making? How are they to be involved? And how are they to become involved? Criteria for choosing revolve around questions involving (1) the right to be involved by virtue of some special capabilities or competences, and (2) expediency, that is, the consequences of recognizing or ignoring claims for involvement. Here, one's role and expectations of self and others, engendered and sustained by that role, critically influence the preferences and prejudices one brings to the task of choosing who should participate.

Probably the first criterion applied would be entitlement to participation by virtue of competence. Competence as a criterion is clear enough when the issue is technical or scientific competence. However, what would constitute competent community participation? Usually, it is assumed that whoever represents the community should be competent in the subject matter. Others, however, fear that persons from the community who are competent in the scientific and technical issues are likely to be scientists or engineers who, thereby, are likely to weigh their judgments by the same criteria as the more directly involved scientists. From their perspective, there are equally relevant competences such as the ability to sense and express the fear, hope, and confusion of laypersons, all of which are claimed to be data as cogent for decision making as technical facts. Yet community representatives—or otherwise participatory community members—must understand the scientific-technical issues well enough to appreciate arguments pro and con for the research. How to provide both kinds of competence is a central and unresolved problem, though the growing capability of consumer information and action groups suggests the challenge is not insuperable.[12]

Another kind of competence belongs to those with formal organizational responsibility and associated skills. (Such competences are represented at the University of Michigan by the regents, laboratory directors, researchers, certain deans, the vice president for research, Committee B, and so on.) Who are the correspondingly competent and responsible members in the community? Community council members? The mayor? Leaders of socially active religious groups? Unofficial but influential groups such as the Ann Arbor Citizens Council or the League of Women Voters? Different groups will differ in their criteria for competence and appropriate responsibility for participation.

Finally, there are the competences needed to represent future generations. Who defines and judges these?

A second general category of claims on participation in decision making relate to the protection of turf. Whether or not there is research will affect the status of persons, the dominance of disciplines, the comparative power of administrators, research directors, deans, and so on. At the University of Michigan, for example, some felt that important contributions to the university's prestige and, therefore, to its future overall research budget, its turf vis-à-vis other universities, would depend on vigorous involvement in recombinant DNA research. Others argued that the university would gain prestige by leading the way in rejecting the research. If the community were to be involved in the decision making, analogous concerns with turf protection would also arise.

Related claims to participation in decision making would be in terms of risk to personal reputation and income—including consulting fees—if the research were not done and for others from physical exposure to risk to these synthetic biological entities if it were done. And if accidental leaks from the laboratory do cause damage, who would be at risk financially if the university is sued by all those allegedly harmed?

All claims on the right to participate would be influenced by the intent of the decisions to be made. That is, what is to be the purpose of the decision? What is its scope? How inclusive is it to be? Is the decision chiefly scientific, or is it political, ethical, or operational? Is it to be advisory or binding? Note, too, that the initial intent for decision making depends on the perspectives and interests of those who decide who is to decide.

This sample of claims to entitlement in the decision-making process emphasizes that claims will be a function of the roles of those who put them forward and that different values and norms will be involved, including many that extend well beyond issues of scientific and technical competence.

I turn now from the question of what criteria would be involved in deciding who is to be involved to the question of what the steps must be for effective community involvement to take place—assuming community involvement in some form. What must happen preceding the decision making in order that it can be accomplished? In this kind of situation, claims on participation are novel, even radical. Participation raises profound problems for the conduct of free inquiry, and those outside the conventional decision-making network who claim a right to participate do so on the basis of values and norms not necessarily compatible with those of the conventional system. So, the usually routine steps of a conventional decision-making system now become serious questions of procedures and tactics. Getting from one point to the next will require new social inventions and probably new norms to legitimate those inventions. These, then, seem to be the steps necessary to set in motion a process for inventing new decision processes.

Steps to the Invention of a New Decision Process

Step 1. The community must organize itself to make its claim for participation. Not only is there the task of generating and focusing community interest, but the question must be answered, Who constitutes community? Crudely, how many people in the community and which groups need to be engaged for them to (1) claim successfully to represent the community interest in whether research should be undertaken, and (2) get the organization that would do the research to accept them as representing that interest? To put the question another way, is it possible to deal with this situation through some agreed-upon community process rather than rely on the makeshift approaches that tend to be used when a community hastily confronts a situation?[13] Can we reach out tentatively with the intention of learning how to do these things deliberately and in good spirits instead of waiting until anger, confusion, and multiple extraneous interests collude to force a messy and uninformative confrontation?

Step 2. Assume that, one way or another, the community has created a representative entity to engage the research organization in discussion of community participation in the decision-making process. Then, who is to be approached for this purpose—that is, how does the community inform the organization of its intention to participate? Clearly, it must reach persons who take seriously the community interest, at least as a matter of public relations, if not out of recognition of the organization's ethical obligation to the community. What is more, those approached must have enough clout to converge and to hold the organization's attention to the question of community participation in decisions previously the prerogative of the organization.

At the University of Michigan, individuals, ad hoc groups (especially the one formed around Dr. Susan Wright's memorandum to Committee B [see note 1]), and the University Values Program all espoused the need to face the question of community participation, but not representing the Ann Arbor community as such, they sought out members of the Board of Regents, the Vice President for Research, members of Committees B and A, the Senate Assembly, and the Senate Advisory Committee on University Affairs (SACUA). The result was university-wide moral and financial support for the forums, including the cost of outside experts. Further conversations, memoranda, and the forums also contributed to the extraordinary attention the regents devoted to the issue.

These activities were influenced by some in the university who worked hard to alert others in and out of the university to the need for a public airing.[14] Their interpretations of the situation converged in a belief that these activities offered real potential for new university-community interaction. But how might this have gone if it had been evident that the Ann Arbor community was going to insist on an active role in the initial decision? That never happened, nor was it expected to when groups in the university agreed to support the forums. It may be that some justified complacency allowed the university to be more innovative than if the community had been more assertive. Thus the absence of crisis made it possible to draw many university personnel into the issue in ways that taught which activities might be useful in less tranquil circumstances.

Step 3. How does the community participate in the procedures through which the organization decides if and how the community is to be involved in the decision process? Intensification of a recombinant DNA type of issue could make this a very real question indeed. The same social forces that produce the demand for a broader decision base also produce the demand that decisions about the "if and when" of that wider decision base involve the participation of the wider base. The end of this seemingly infinite regress would appear to lie within the organization, since it is being petitioned by outsiders and since it has the organization and traditions for making decisions concerning the extent to which it is willing to alter its decision process. These decisions depend on beliefs about who can act, in what kind of decision, conducted according to what procedure. They depend, too, on the process by which the organization would arrive at a decision that participation is permissible. And this process in turn depends on the accepted definitions of competence and turf described earlier.

However, not all the options lie within the organization: the community could seek legal redress, in which case the decision about who has a right to be involved might be made outside the organization. In the Ann Arbor-University of Michigan situation, the regents played a more active role than usual in deliberating the proposed research. On the basis of their legal obligation to protect the general interests of people of the state vis-à-vis those of the university, the regents might have sought to have persons from Ann Arbor involved in the decision.

Under other more intense circumstances, though highly unlikely, even though the university might steadfastly protest interference with freedom of inquiry, the regents or judicial authority might require the university to include Ann Arbor citizens in the decision-making process, specify the criteria for selection, and determine the decision-making procedures.

There is a less precipitous approach to the question posed in step 3. Collaboration and inventiveness depend on the extent to which trust can be built up between the interested parties and, through trust, appropriate norms evolved. Denial of organizational legitimacy and insistence on fuller participation are in part the result of well-documented, widespread distrust of the conventional decision-making processes in large organizations. Trust and shared norms probably can only be reestablished under circumstances that encourage and reward experiments with new decision-making methods and norms that are explicitly designed for decisions about who is to make risk-relevant decisions. The shared experience of learning how to do these things seems prerequisite for creating decision-process norms commensurate with the seriousness of the impact of esoteric and powerful science on an increasingly complex and vulnerable world.

Step 4. Assume that a decision is made to involve citizens in subsequent decisions. How, then, can the decision-making process be operated so that the citizen members can have a truly potent role in influencing outcomes? There is no reason to expect that new procedures will work well initially, even if it is clear what well means—which it certainly isn't. However, a primary condition to be met is insuring that learning will occur that leads to improvements. In Roland Warren's words, "We need to find ways of channeling change which will assure that you and I will reach the optimum agreement possible, but that our remaining disagreement will neither immobilize us nor result in our destroying each other and those around us."[15] Some aspects of decision making that require experiment and learning follow.

How are decisions to be arrived at—by consensus, by a committee, vote, or referendum? By what proportion? How is information to be presented and evaluated? According to an adversarial procedure or collaborative synthesis? Such questions bear on decision-making procedures and on the proportional composition of decision-making groups. Whether decision-making entities in fact set policy and operations or whether they merely make recommendations to other entities would be additional considerations. And anticipation of how these considerations will be dealt with will influence actions associated with steps 1 through 3.

At this stage proposals come into play for combining technical and social considerations in decision making for public policy. These include such proposals as the science court,[16] judgment analysis,[17] and judicial evaluation.[18] Intriguing, hopeful, and controversial as these are, they do not of themselves vitiate the new and difficult task of getting to the stage where they can be tried out. Their use implicitly assumes that decisions have already been made about who will make the decisions facilitated by these procedures. However, explicit plans to experiment with such procedures might simplify questions regarding appropriate competence and the intent of the decision making. At least such plans could color expectations about what is to be done and how. This, in turn, could contribute to the building of trust and shared norms.

Protecting the Effectiveness of Minority Viewpoints

Community members may find themselves in a minority in the decision-making entity if their role leads them to emphasize different perspectives that reject or question the conventional wisdom of the experts about the costs and benefits of proposed research or the appropriate context for evaluating them. Then, too, they may be a minority in numbers as well as in their position on the issue. Especially because of the possibility of different interests among the community members, minority positions must have access to special resources in order that they may make the best cause they can as they develop their position and so they can disseminate it to all potentially supportive constituencies. In those novel and momentous areas involving powerful new scientific knowledge and technique, the minority position may be the one that merits the most intensive, early amplification if wise decisions are to be made later.

Minority positions, then, must be able to command:

1. Access to sufficient information and resources where the information can be developed into the best case they can make. Technical understanding may need augmentation and those needing it should have access to additional information sources. Sometimes such ac-

cess would depend on funds to undertake alternative technology assessments and/or social or environmental impact studies.

2. Sufficient "presence." A devil's advocate will not be enough. Though good for the conscience, the role is usually insufficient for effective influence.[19]
3. Sufficient resources to disseminate broadly their position so that others who might find it attractive will learn about it. Typically, minority positions lack both dissemination capabilities and legitimacy. Therefore, part of the responsibility of a decision-making entity ultimately responsible to the public interest should be to ensure that its minority positions have both.

In Conclusion

One fact is clear in all of the swirling ambiguity of positions and counterpositions about the state of society and what needs to be done about it. We are too ignorant of our own condition and its potentialities and problems to engineer our way into the future either materially or socially—we cannot get there the way we got to the moon. Instead, we must learn to create a new set of norms, values, and supporting behaviors that will allow us to continue to be a learning society in terms of where we think we are, where we think we want to go, if we are getting there, and if we still want to. Rapidly changing circumstances permit no other mode of rational conduct.

Changing values, risks, and ambiguities require profound, perhaps radical, changes in the norms by which decision-making entities in research-oriented organizations operate and in their ends. These entities also must learn how to design themselves so that they are effective learning systems to improve the effectiveness of community and organization participation in decisions about scientific activities that contain potential community risks.

More specifically, this requires decision-making entities to learn their way through new issues wherein the public interest seems to confront freedom of inquiry:

1. A shared learning relationship instead of an adversarial stance. A zero-sum approach, an assumption that there is one right answer and that only one side can win, can only lead to disaster.
2. An openness to continuous reexamination of the norms and values by and for which they operate. It will be especially necessary to reexamine continuously the means for estimating and evaluating social costs and benefits. Alternate scenarios will be needed so that the community and the research organization will have the broadest possible perspective for decision making in these ambiguous and ambivalent areas.

3. Effort, money, and attention devoted to learning how to learn in these situations, to learning how to integrate persons, ideas, and actions based on new normative modes.

Whatever decision-making entities decide and however they do it, experiment, that is, research and development on the norms and processes of decision making wil be unavoidable.

Surely this sounds utopian, yet as Bertrand Russell observed, a utopian perspective is the only practical one, and this is especially true in the kind of world exemplified by the recombinant DNA issue. In learning how to make public decisions involving potentially risky esoteric research, we must commit the same kind of intensity of imagination, experiment, and time to learning how to conduct decision-making processes as we have to learning about nature. If we do, then we can hope that, even though a particiular mode of participation or outcome may not satisfy everyone, the norms developed in arriving at them will be satisfying enough to sustain a sense of community while other processes evolve. For some of us, the University of Michigan experience was a beginning of the kind of learning that could help realize that hope.

Notes

1. It is right that at the outset I give my personal position on the topic of this chapter. In November 1975, Dr. Susan Wright, a lecturer in the history of science at the Residential College, University of Michigan, shared with me and a few others a memorandum she was addressing to Committee B. (Committee B was a faculty member group established in 1975 by the Office of the Vice President of Research to determine university procedures for approving recombinant DNA research.) In her memo, Dr. Wright requested that more attention be given to certain aspects of the risks associated with recombinant DNA research. (Until then I was unaware of Committee B or of the question of recombinant DNA research at the University of Michigan.) I became involved in efforts to bring the community into the picture through participation in a small ad hoc group inspired by Dr. Wright's concerns, as a member of the group guiding the University Values Program, and later, as a member of the committee designated to plan the university-community discussion forums described in this chapter.

   I became involved because of my concern with the issue per se and because the recombinant DNA research issue was an invaluable occasion for the university and the community to begin to learn how to deal with such issues. My personal (cautious) inclination is to favor community involvement in the basic decisions. Cautious because I also acknowledge the dilemmas and difficulties described in this chapter.

   It remains for me to acknowledge that we who ponder on and seek to act regarding the place of science in society are caught in a maze of distorting mirrors that reflect the currents and conflicts in our culture and its many subcultures and, therefore, in ourselves. We too are mirrors caught up in the maze and contributing to the maze. No matter how much we act with good will and seek to be unbiased, we are, ineluctably, mirrors.

2. See Leon R. Kass, "The New Biology: What Price Relieving Man's Estate?" *Science* 174 (19 November 1971): 779–88.
3. "Environmental science, today, is unable to match the needs of society for definitive information, predictive capability, and the analysis of environmental system as systems. Because existing data and current theoretical models are inadequate, environmental science remains unable in virtually all areas of application to offer more than qualitative interpretations or suggestions of environmental change that may occur in response to specific actions." National Science Board/National Science Foundation, *Environmental Science* (Washington, D.C.: National Science Foundation, 1971), viii.
4. Recall that two large commercial aircraft collided over the Grand Canyon and that another two collided over New York City. An Air Force bomber hit the top floors of the Empire State Building. The ocean liner, *Andrea Doria*, sank after a collision with another ocean liner in clear, calm weather in mid-ocean. The oil tanker, *Torrey Canyon*, went aground on well-known shoals spilling oil all over the southeastern English coast. The unsinkable *Titanic* sank on its maiden voyage.
5. See Todd R. LaPorte and Daniel Metlay, "Technology Observed: Attitudes of a Wary Public," *Science* 188 (11 April 1975): 121–27.
6. See William Bevan, "The Sound of the Wind That's Blowing," *American Psychologist* 31, no. 7 (July 1976): 481–91.
7. See the sophisticated statements pro and con limiting freedom of inquiry found in Hans Jonas, "Freedom of Scientific Inquiry and the Public Interest," *The Hastings Center Report* 6, no. 4 (August 1976): 15–19; and R. L. Sinsheimer and G. Piel, "Inquiring into Inquiry: Two Opposing Views," *The Hastings Center Report* 6, no. 4 (August 1976): 18–19. For other straws in the wind see Barbara Culliton, "Kennedy Hearings: Year-long Probe of Biomedical Research Begins," *Science* 1973 (2 July 1976): 32–36; and T. Seay, "Stoned in Peoria," *APA Monitor*, June 1976, pp. 11–12. The latter article is about Congress's refusal to fund research already approved by the National Institute on Drug Abuse.
8. Notes 2 and 3 are also relevant here.
9. This extraordinary and laudable social invention is itself evidence of the changing norms in science with regard to social responsibility. It certainly merits systematic study—which it hasn't gotten—for the deeper understanding it could provide about the social and psychological conflicts and clarities unfolding to today's science community.
10. That there was acknowledged concern about the possibility of arrogant disregard of overly stringent guidelines is evidence of another aspect of the normative and ethical disarray of this society—the same society that engendered the voluntary moratorium on recombinant DNA research.
11. Apparently the anticipated risk was perceived as too small to justify the complexities and delays associated with serious examination of the possibility of re-

stricting the chances of accidents to regional or national laboratories analogous to Brookhaven, Argonne, NIH in-house research, or the great multinational research installation, CERN. Such facilities, if located well away from dense human habitats, would eliminate the issue of who locally is entitled to participate in decisions.

At the University of Michigan, some of us, especially Professor Max Heirich, urged the regents of the university to join with their counterparts at other involved universities to seek funds from the federal government for a jointly shared, isolated laboratory. Such an effort by the regents would have been unprecedented and time consuming. But some of us urged that circumstances merited such a social invention from the group bridging the university to the larger community—the regents being elected by the public at large. The regents did not act on this recommendation.

12. There are growing numbers of consumerist organizations, think tanks and consultants that can provide such information and knowledgeable spokespersons. Scientists and engineers are prominent resources in most of these groups. Examples are Science for the People, Federation of American Scientists, Scientist's Institute for Public Information, Global Tomorrow Coalition, various offspring of Ralph Nader's groups, and ad hoc groups such as those arguing against nuclear reactors.

13. A dilemma: how to make the community aware that there is a risk and that the consequences may be grave without inflating the issue to panic proportions. Panic would obviate deliberate and enlightened decision making and also destroy chances for emergence of an attitude that would make the eventual decision at least tolerable to most if not all parties. While not precisely analogous, the response of Cambridge, Massachusets, to Harvard's research intentions is most informative. See, for example, "Recombinant DNA: Cambridge City Council Votes Moratorium," *Science* 193 (23 July 1976): 300–01.

    A related difficulty merits comment. If the research organization is a university, the chances are—as in the cases at the University of Michigan and Harvard—that some community interest will be stimulated by university personnel. While it needn't work this way, it is likely that signals of concern, especially the early ones, will be carried from the university to the community by university people. If community interest grows and if that interest is antagonistic to the conduct of the proposed research, the risk of polarization within the university itself will also grow. Polarization would destroy the openness necessary among university members if there is to be social learning and invention of the high order that will be required to cope with such problems.

14. Committee B had been open to input from the community, but until the aforementioned groups became active some two months before the forums, chiefly as a result of Professor Susan Wright's memoranda to Committee B, there had been little public or university-wide attention to the matter.

15. Roland Warren, *Love, Truth, and Social Change* (Chicago: Rand McNally, 1971), p.298.

16. See Task Force of the Presidential Advisory Group on Anticipated Advances in Science and Technology, "The Science Court Experiment: An Interim Report,"

Science 193 (20 August 1976): 653–56; and Philip M. Boffey, "Science Court: High Officials Back Test of Controversial Concept," *Science* 194 (8 October 1976): 167–69.

17. See R. L. Wolf, J. Potter, and B. Baxter, "The Judicial Approach to Educational Evaluation," April 1976, a transcript of an instructional tape on the Judicial Evaluation Model. The tape was presented at the Annual Meeting of the American Educational Research Association, San Francisco, April 1976. Information about this tape can be obtained from Dr. Robert L. Wolf, Education 325, Indiana University, Bloomington, IN 47401.
18. See K. R. Hammond and L. Adleman, "Science, Values, and Human Judgment," *Science* 194 (22 October 1976): 389-96.
19. For a most perceptive critique of the devil's advocate role and its limits, see A. Hirschman, *Exit, Voice, and Loyalty* (Cambridge, Mass.: Harvard University Press, 1970).

The various combatants in the DNA war could be grouped, somewhat loosely, into several identifiable entities with certain common characteristics. That is, there were "the scientists"—not all scientists, of course, but the majority of molecular biologists who argued that the risks of gene splicing were minimal and that the federal regulations or legislation (except as necessary to override state and local rules) were unnecessary. Then there were the scientific professional organizations, generally representing the interests of scientists but taking a much more diplomatic, politically astute, and professional approach than their member scientists. Perhaps the Congress might be considered a third entity, although within this group were to be found the champions of any of the other factions.

One of the most significant among the various coalitions was clearly that of the public interest groups. Although they were not of a single mind on all issues, there were some important points of agreement. They were unanimous in contending that local communities should have the right, as an expression of popular democracy, to impose whatever rules they wished on scientific research and technology, even if they were much more stringent that the NIH guidelines and had no particular scientific basis. Insufficient knowledge of harmful consequences, they argued, was reason enough for citizens to exercise their democratic rights. In large part because of the lobbying of these groups, and with the support of their populist congressional allies (such as Edward Kennedy, Barbara Mikulski, Henry Waxman, Edward Markey, and Andrew Maguire), the matter of federal preemption of local laws became the most contentious issue of the debate as it was fought in Congress.

Claire Nader has been doing research on the interface of science, government, and the public for many years. She has worked as a social scientist at the Oak Ridge National Laboratory and later extended her investigations to Washington, D.C., where these three spheres intersect. As the contribution of Norton Zinder is revealing in capturing the essence of how the controversy was perceived by the scientific community, Dr. Nader's essay is an expression of the spirit of the public interest world view.

9

Technology and Democratic Control: The Case of Recombinant DNA

CLAIRE NADER

Introduction

The controversy over recombinant DNA research highlights the tension between the seeming imperative of science-technology and the values of a democratic society. Do the techniques and technologies that scientists and engineers develop advance or subvert public purposes and values? This central question underlies critical socio-technical controversies and is discussed here in relation to definition of issues, the concept of countervailing force, and arguments over mandatory standards.

The first section of this chapter reviews features that characterize public interest movements in general and, by comparison, those that characterize the promoters of emerging technologies. The discussion draws on nuclear safety and similar issues to suggest how citizen involvement can widen the context for public debate and public decision making on recombinant DNA technology. The second section, also utilizing the experience with other technologies, highlights the edge that promoters of rDNA have over its critics and over ensuing developments in the absence of a sustained countervailing force. The third section considers the social purposes that mandatory standards can serve, what is lost if standards are abandoned, and what happens when probability specialists dominate consequence specialists in technology development in general and in the politics of the gene-splicing enterprise in particular.

There is a class of technologies that can do great, perhaps irreversible harm. Recombinant DNA is a member of that class. Its life-shaping properties demand, as biochemist Liebe F. Cavalieri has observed, "a new consciousness if human life on this planet is to continue."[1] A new consciousness requires broad citizen involvement. But in order for citizens to decide whether or not their society is equipped to manage the creation of new life forms, their rights to know and participate must be satisfied. These rights, and their relationship to remedies, are discussed in the fourth and final section of this chapter.

Defining the Issues

By the time the controversy over recombinant DNA research surfaced in the middle of the 1970s, the shape of today's public interest advocacy had already acquired a definable form. The movement had cut its teeth on a varied agenda of public problems, including pesticides, unsafe automobiles, harmful uses of medical diagnostic x-rays, contaminated meat, dangerous drugs, environmental pollution, hazardous household products, and the risks of nuclear power.

When debates on such issues develop, they generally reflect existing power structures, including those in science and technology. The drive behind the development of many important technical applications in the United States is supplied by networks of staunch advocates from the science and engineering fields and the government, academic, and corporate communities, often aided by enthusiastic and uncritical reports in the media about their views and programs. What is not reported, and may not even be known without considerable investigation, is the interlocking agendas of these various powerful communities and what these agendas mean for society in critical areas of life (for example, in agriculture, health, communications, banking, energy, and defense).

Many technologies with substantial social consequences get off the ground with little or no public discussion, even though considerable amounts of tax money may have supported the take-off. More than twenty years passed, for example, before nuclear power technology—almost totally financed with tax dollars—came under sustained public scrutiny. Another dramatic example is the arms race. Here inventors, giant companies competing for defense contracts, interservice rivalries, federal officials advancing their status, and members of Congress with various interests to protect are all a part of an influential lobby for the arms technologies. Thus an arsenal of destructive weapons has been built with almost no public involvement in deciding how best to insure the nation's security.[2] Citizen activity in arms control questions has recently developed but has nowhere near the structure and power of the alliance for these ever more dangerous weapons.

Citizens are up against powerful adversaries when they criticize any well-supported technology that can control the public's access to knowledge about it. Moreover, alliances in favor of a technology do not disband easily even if serious problems surface and the public is aroused. Too many special interests are at stake; too many career commitments are entrenched. A prominent nuclear power advocate illustrated the problem when asked to revise his long-held position in view of the safety problems about which he expressed private concerns: "But you are asking me to say my whole career has been wrong."

Technological proponents typically share an uncritical enthusiasm for the technology; they focus on its benefits, underplaying negative consequences; and they have a firm belief in "tech fixes" and the primacy of experts and decision making by the few who "do know." These characteristics lead them to define the public good in terms of narrowly conceived private interests (including profits, careers, and ideologies). Proponents find a final source of comfort, when something does go wrong, in blaming the users, operators, or victims.

In the automobile industry, for example, for years the focus was on driver behavior rather than on automobile design to reduce highway deaths and injuries—in essence, the victims were blamed for the accident and injury statistics. In the chemical industry, the problem is said not to be chemical company policies and the uncontained toxic chemicals in the workplace, but rather the workers who allege that they get sick from the chemicals. In the international arena, the problem is not nuclear weapons and advocates of an ever-growing arsenal that destabilize superpower relations, but those who favor a freeze on these weapons by both superpowers. In the nuclear power industry, the problem is said not to be nuclear reactors and their unrelenting advocates, but the environmentalists who obstruct the licensing process with questions about safety, or citizens who just do not understand that the probability of a serious accident is small.[3] Citizen activists who have questioned the health and safety aspects of significant technologies, in case after case, have found themselves up against a powerful, entrenched oligarchy that challenges the prospect of democratic governance, the freedom of citizens to choose, and the prospect of a more equitable distribution of political and economic power.[4]

Recent public interest activity is marked by a continuity of concerns. The first of these is to understand the political and economic forces that may insulate a given technology from public assessments of hazards. A second is how the technology will affect people's health and safety now and in future generations. A third concern is that there be accountability by the technology's protagonists to those people who stand to lose the most when things go wrong. A fourth concern is the perception that involved specialists are unlikely to be able to pay systematic attention to harmful or potentially harmful consequences. Another related concern is the ethical questions that arise when specialists, hampered by their limited analytical concepts and techniques and/or their client allegiances, produce incomplete or otherwise misleading assessments of the consequences of the technology.

The public interest approach typically has taken a broad ecological perspective. Citizen appraisals of nuclear power technology, for example, moved from an initial concern with thermal pollution to radioactive emissions to the whole nuclear fuel cycle and how this affected workers and the environment—from the mining and enrichment of uranium to the siting, construction, and operation of the reactors, to the cost of nuclear-generated electricity, to the transportation and disposal of radioactive waste, and to the grossly inadequate emergency evacuation plans of nearby communities in the event of a nuclear accident. Citizen groups also focused on the fraternity that had nourished the drive for nuclear power since World War II, including technical specialists, government bureaucrats, congressional committees, manufacturers of nuclear reactors, and electrical utilities that were major buyers of reactors. As public learning progressed, technical, environmental, health and safety, political, economic, and ethical questions as well as generational consequences all became part of the analysis.

For their part, nuclear power advocates were neither uninformed nor unconcerned about safety problems. Government officials undertook numerous safety studies (which citizen groups then obtained under the Freedom of Information Act and used to build their own case), and nuclear safety questions were debated—but only with "insiders." When a problem proved intractable, they clung to the technology and hoped for a tech fix. Lack of a solution for the safe disposal of radioactive wastes, for example, did not lead to deferring deployment of nuclear reactors pending a solution. There were even some nuclear entrepreneurs who comforted themselves with the argument that the country already had an enormous burden of radioactive waste from its nuclear weapons program. The genie was out of the bottle, they would point out, and what power reactors added was miniscule. And such positions were easy to take as long as the decision process was closed to "outsiders."

Public interest advocacy groups hold firm to the proposition that the public interest is served best by an open discussion and open decision making in institutions that govern or profit from a risky technology. Such institutions should not make life-shaping decisions in secret. Rather, they have an obligation to understand the hazards and tell the public. In turn, the people have an obligation to question so that finally "what touches all is decided by all," as the Roman adage put it. Moreover, openness both in government and in corporate decision making on health and safety matters is a version of the marketplace of ideas and provides for the development of the best course of action and the fewest illusions for all. With the growing *public* complexion of the debate on nuclear power, for example, the issues were joined in a wider context of values than that held by nuclear advocates—an essential benefit of the democratic process.[5]

Public interest advocacy, holding that responsible use of corporate and government power requires the mobilized efforts of citizens, finds that the going is not always easy. The first criticisms by outsiders are often dismissed. Controversy may be provoked, but it can readily subside as insiders take advantage of their greater information and monetary resources. And the insiders can take public silence for public acceptance even though lack of public knowledge of a technology's social effects is a more likely reason for the silence. The past decade and a half has shown that acceptance of a variety of industrial technologies has taken a downturn as the public learned about their hazards.

The level of public discourse and citizen involvement in any of these issues is affected by a number of factors, including events, amount of available resources, and timing. The initial controversy may bring forth a vigorous public response, then the issues are quiescent until something causes controversy to erupt again. When the problem is rediscovered, an organized, full-time citizens' advocacy capability may take root. Citizen concern over the issue of nuclear safety has supported such groups since the middle 1960s. Though the resources available are tiny compared to those of the alliance for the nuclear power technology, they have produced documented analyses sustaining citizens' advocacy positions in government hearings and in court.[6]

Public advocacy groups need not and generally do not take purely negative positions. Discovery of a host of problems with nuclear power, for example, led public advocates concerned with the health of the whole to search for alternative energy sources that would be safe, reliable, economical, and renewable, and in addition would lend themselves to decentralized power systems, thus reducing the nation's vulnerability to power failures and increasing self-reliance and self-government. These advocates argued that energy technologies were *means* that should be evaluated in terms of how well they advanced *ends* desirable for the country as a whole.[7] Setting out the problems of technologies in this way turns the debate to normative questions and reminds us that *can* does not necessarily imply *ought*.

The democratic process requires that the development of any technology having a substantial public consequence begin by including a wide-ranging public discussion of social purposes to guide the selection of strategies and operations. Someone must represent the public interest, for it cannot be left to private interests imbued with the attitude—as General Motors President Charles Wilson once suggested—that what is good for General Motors is good for the country. As long ago as 1971, Nobel laureate James Watson cautioned against the same private interest mind-set in science:

> The belief that . . . science always moves forward represents a form of laissez-faire nonsense dismally reminiscent of the credo that American business if left to itself will solve everybody's problems. Just as the success of a corpo-

> rate body in making money need not set the human condition ahead, neither does every scientific advance automatically make our lives more "meaningful". . . .[8]

And physicist Freeman Dyson has commented instructively on how means become ends—how "technical arrogance overcomes people when they see what they can do with their minds."[9]

The priority given to the general welfare in citizen assessments of the consequences of various technologies is sometimes described as just one more special interest pleading. This may be, as John Kenneth Galbraith has suggested, because we have invested more in the private than in the public citizen sphere.[10] And certainly the influence of self-initiated citizen advocates does not stem from any political or economic power. Rather, it is an exercise of First Amendment rights, and public interest arguments must be accepted or rejected on merit because there is nothing else to attract support for their positions. Nevertheless, from earliest days this country's citizens have taken noteworthy initiatives to decide the issues that affect them deeply.[11] Public interest advocates in communities and from the ranks of government, universities, labor, and corporations have continued to speak out on problems according to their consciences,[12] including the biohazard problem and other aspects of gene-splicing technology.

It was against this background of an advocacy dedicated to citizen involvement that public interest organizations engaged in the rDNA controversy. They brought a set of systematic concerns to it, both substantive and procedural, and moved this issue from one that molecular biologists were attempting to resolve, mostly among themselves, into a full-blown public debate.

Developing a Countervailing Force

Recominant DNA techniques provide new DNA hybrids by combining parts of DNA molecules from two difference species. The new hybrid DNA is introduced into selected bacteria that can then be used to produce large amounts of substances such as hormones or other physiologically important molecules (see Appendix). Among the suggested or already achieved medical benefits are the production of human insulin to replace the animal insulin now in wide use, human growth hormone, interferon for treating several kinds of cancer, and human and animal vaccines. A growing understanding of genes may make possible the correction of genetic defects in humans. Genetically modified varieties of grains and other food plants promising higher yields and higher nutritional value are being produced, and other projected agricultural benefits include nitrogen fixation in plants to increase yield, cut costs, and reduce dependence on expensive energy used in synthetic fertilizer production.

Unfortunately, every expected benefit in genetic manipulation carries its own risk,[13] and critics who worry about how little we know point this out. Our vast ignorance, wrote biologist Lewis Thomas, is "most of all about the enormous, imponderable system of life in which we are embedded as working parts. We do not really understand nature at all."[14]

Risks take two forms: those that come in the doing of basic research and those that may accompany applications.[15] The possibility that harmless bacteria could be transformed so as to produce serious diseases in humans—diptheria, for example, or cholera—is an example of risk in basic research. The risks that accompany applications are harder to specify because applications are not fully known. However, the genetic manipulation of microorganisms using recombinant DNA technology is now widespread, and workers in laboratories and factories are inevitably at risk.[16] We can expect that some of these microorganisms will escape into the ecosphere. There will also be the intentional release of industrially useful microorganisms, for example, those designed to digest oil spills (will they then proceed to eat oil wherever it is found?) or to make cellulase, a family of enzymes that can degrade cellulose (including in the human gut?). In another area, what could be the result of solubilizing toxic trace metals in mining operations? Unintended effects are precisely what continue to worry critics of present recombinant DNA technology.[17]

Researchers themselves first expressed concern about potential biohazards of recombinant research in a letter published simultaneously in Great Britain and the United States in July 1974.[18] The initial experiments conducted to answer certain questions about possible biohazards quieted the fears of some scientists, but their concern created controversy on university campuses and in communities across the country during 1976 and 1977.[19] Molecular biologists, local government officials, and ordinary people all became involved in illuminating exchanges. The most celebrated took place in Cambridge, Massachusetts, where the blunt and colorful Mayor Alfred Vellucci articulated a main issue in the science-society relationship:

> It's about time the scientists began to throw all their god-damned shit right out on the table so that we can discuss it. . . . Who the hell do the scientists think they are that they can take federal tax dollars that are coming out of our tax returns and do research work that we then cannot come in and question?[20]

Several months later in November 1976 and five months after the National Institutes of Health (NIH) had issued its first guidelines for rDNA experiments (see earlier chapters in this volume), citizen groups responded formally. The Environmental Defense Fund and the Natural Resources Defense Council petitioned the Department of Health, Education, and Welfare for a public hearing to consider whether rDNA research should be allowed, and, if allowed, what conditions researchers would have to meet. The petition aimed at "a broad-based public review of the existing NIH guidelines" to expand the "debate on issues given little attention by the NIH drafting committee of the office of the director," such as the appropriateness of *E. coli*—commonly found in the human gut—as an experimental host. In January 1977, the Sierra Club Board of Directors proposed putting all rDNA work in maximum containment laboratories, which the federal government operated or controlled, until more information was available and openly discussed. In March 1977, the Friends of the Earth called for a moratorium, also until a more thorough investigation had occurred.

The Coalition for Responsible Genetic Research—a group of specialists and other concerned citizens formed in the mid-1970s—worked to maintain effective NIH laboratory safety guidelines. Its members wrote articles, spoke before scientific and environmental meetings, organized conferences, provided technical assistance to community and labor groups, and drafted local ordinances.

The public questioning grew apace. In the legislative arena the controversy sharpened enough for the proponents to feel the heat of opposition. The critics of gene-splicing work sought mandatory federal regulations to cover both academic and corporate researchers and pushed for mechanisms to insure full assessments and a policy voice for citizens. But legislative activity ended without resolution (see Chapter 3). The powerful standing of scientists who favored genetic engineering, including those scientists who organized into a lobbying force in support of NIH stewardship (see Chapter 5), the insufficient leverage of critics, and the futuristic issue's low priority for many members of Congress (no accidents had occurred) were some of the factors which helped deflect congressional attention.

The arena for public discourse then shifted back to the National Institutes of Health's Recombinant Advisory Committee (RAC) and its work on rDNA guidelines. Groups that had been active in the legislative debate went to its meetings also, although in recent years limited budgets, shortage of personnel, and other constraining factors have limited participation of the citizen groups. While industrial and government observers continue to attend regularly, only a few investigators, such as Susan Wright, a University of Michigan historian of science who attends the NIH-RAC meetings as part of her research comparing the U.S. and British handling of genetic engineering work,[21] continue to observe and raise important health and safety questions. (Some members of RAC also had concerns about health and safety questions, but their voices were not the dominant ones, as events have shown.[22])

From time to time the Coalition for Responsible Genetic Research has issued statements of concern—for example, one on the precedent set by the June 1980 Supreme Court decision allowing the patenting of man-made living organisms and another, in a letter to the NIH dated 29 July 1981, opposing a proposal to abolish completely the NIH guidelines and oversight functions.[23]

Gene transplantation may have been the first innovation to be debated publicly before the widespread use of the techniques and before large investments in applications had given it economic power that was hard to oppose, as one commentator noted.[24] Even so, a full-fledged citizen capability to monitor the genetic engineering has yet to emerge. And until a sustained countervailing force compels their resolution, or evolution, major issues will continue simply to simmer. At the same time, a burgeoning biotechnology industry, in a budding and close alliance with universities, is now creating a new set of conditions not only for public policymaking but also for the conduct of science.[25]

These conditions include a definite trend toward secretiveness that the drive to abolish the government guidelines and the Supreme Court decision reinforce. Corporations thrive on secrecy, assertedly in the name of competition. As commercialization of rDNA technology accelerates, corporate trade secrecy will dominate even further the tradition of open communications in science. "Scientists who once shared prepublication information freely and exchanged cell lines without hesitation are now much more reluctant to do so," noted Stanford University president Donald Kennedy several years ago. Commercialization, in his view, also threatens graduate student access to significant molecular biology information.[26]

Science, controversy, and dissent depend on openness. But the industry-university connection complicates free speech and free access. As Harvard University scientist John Edsall has commented, "The techniques of gene cloning hold forth the promise of manufacturing substances of great biological importance, cheaply and on a large scale."[27] This promise beckons the research scientist who wants part of that economic action, as both university professor and company executive, as holder of substantial stock options, or as a major consultant to bioengineering companies. There is the well-known combination that David Baltimore has fashioned. He is a professor at MIT; board member and major equity holder in Collaborative Research, a biotechnology company; and director of the recently established and controversial Whitehead Institute which was created outside the university to do recombinant research. This novel arrangement provoked a stormy exchange at MIT over the terms that would define the MIT/Whitehead relationship.[28]

Under such circumstances where will loyalties lie—in the educational and scientific enterprise, or in the corporate enterprise? Furthermore, if educational and research choices are commercially driven, whence will come well-grounded technical assessments of genetic engineering developments?

There is a power imbalance among scientists also. Those who advocated the use of the gene-splicing technique *before* the fundamental questions of their worried colleagues were answered remain on the offensive. A long-time chronicler of the rDNA debate, Charles Weiner, has observed that the 1975 Asilomar conference on biohazards itself served to swell the ranks of proponents, attracting scientists not before involved. Moreover, there was a conspicuous absence at Asilomar of specialists in disease prevention and public health, and specialists to explain how political and economic forces can thrust a technology forward leaving those who want to proceed cautiously striving just to keep things from getting completely out of hand. Had these specialists been represented, perhaps a broader, more stable alliance for risk questions, health and safety standards for the experiments, and ethical considerations could have been developed.

Now, almost a decade since the first NIH guidelines were issued, the constituency for risk management and strict standards remains fragile. Since 1976, major and minor revisions at each meeting of the Recombinant Advisory Committee have virtually insured the ultimate demise of these guidelines.[29] The trend has been twofold: one widening the type of donor and host organisms now in use; the other weakening containment requirements and oversight mechanisms. Thus, by December 1981, about 85 to 90 percent of the work at universities has been exempted from NIH guidelines. In addition, the responsible institutional biohazard committees function at different levels of commitment and performance.[30]

The attempt to abolish the mandatory nature of the government's safety guidelines as well as eliminate the Institutional Biosafety Committees (IBCs) prompted three members of Congress to protest in a letter dated 20 November 1981 to the chairman of RAC. Although the 1 April 1981 proposal of David Baltimore of MIT and Alan Campbell of Stanford University changing the guidelines into a voluntary code of practice did not pass at the February 1982 meeting of the RAC, an alternate plan by Susan Gottesman, a senior molecular biologist with the National Cancer Institute, did. That plan kept the guidelines and restored the IBCc but dealt a blow to oversight and containment provisions. "The Gottesman proposal," concluded Susan Wright, "stopped just short of the final coup de grâce which the Baltimore-Campbell proposal would have delivered to the NIH controls."[31] RAC recommended the Gottesman plan by a vote of seventeen to three, and the acting director of NIH approved it on 21 April 1982.

What is the rationale for abolishing the NIH guidelines? From the beginning scientists promoting rDNA research have exhibited an antipathy for any outside controls. At Asilomar it was apparently not the cautions of scientists such as Sinsheimer, Chargaff, Beckwith, King, or Singer nor those of Andrew Lewis who warned about the dangers of using SV40 and other animal tumor viruses in recombinant DNA experiments, that persuaded their enthusiastic colleagues of the wisdom of standards. Rather, it was the lawyers at the Asilomar conference who convinced the scientists of the need for guidelines. The lawyers simply explained the public's recourse under liability law should an accident occur.

Indeed, even the use of the word "guidelines" instead of "standards" is another indication of scientists' aversion to outside restrictions. Guidelines are usually voluntary, whereas government standards are mandatory and carry penalties. Even though the guidelines for rDNA are mandatory for government-funded researchers, the penalties are minimal. The worst penalty seems to be the withdrawal of a violator's funds.

Not all scientists were so confident about a future without risks. For example, at the September 1981 meeting of RAC, member Richard Goldstein of Harvard University was "not convinced there aren't any more risks" so that it made sense to eliminate the government's NIH guidelines. He noted the possible hazard associated with reproducing the genes for making botulism toxin and other biological poisons in microbes that have never been able to make such poisons.[32] This once prohibited work can now be done with the approval of RAC and NIH, a legacy of the Gottesman proposal.

Other possible hazards also argue for strict controls. The occupational hazards may be at the front line. Large-scale production of microorganisms might expose workers to dense aerosols of microorganisms in the process of fermentation and processing, with unforeseen results. Or genetically engineered organisms in the laboratory and factory may affect workers adversely even if these could not survive long enough to affect those outside of the workplace.[33]

The evidence to date leaves significant questions about hazards unresolved; experiments are needed to test specific hazards. And the status of rDNA hazards and controls makes a compelling case for stronger, not weaker, mandatory standards.

The issue of public control is again on the Congressional burner. Representative Albert Gore, Jr. held hearings in June 1983 on the environmental and ethical questions associated with genetic engineering. Among Gore's interests has been the creation of a federal advisory commission to track genetic experiments on people. However, legislation to this end has not yet passed.[34] The Environmental Protection Agency is moving to play an active part in reviewing proposals to conduct field tests of genetically changed organisms, covering researchers from industry and academia.[35] On 31 December, 1984, the Office of Science and Technology Policy, Executive Office of the President, published in the *Federal Register* its *Proposal for a Coordinated Framework for Regulation of Biotechnology* and called for public comment.

In September 1983 several environmental groups brought a lawsuit against the NIH over the issue of intentional release of genetically modified bacteria into the environment without first preparing an environmental impact statement in accordance with the National Environmental Policy Act.[36]

While the erosion of the guidelines continues, corporate activity is intensifying, with projections that in ten to fifteen years the genetic engineering business will be a $10 billion industry.[37] Over 150 firms are now in the business of designing microorganisms for industrial uses. Past experience tells us that in the absence of strong, full-time public involvement and adequate government health and safety standards, corporate interests will prevail over the general interests of the citizenry. The result will be eroding public confidence that government at a distance can protect people from unnecessary health risks. One sign of this is the local ordinances by which communities faced with biotechnology companies in their area try to protect themselves.[38]

Citizen action sometimes draws public attention. At other times, between the accidents, lawsuits, legislative and regulatory activity, demonstrations, or sit-ins—the dramatic events that grip national attention and shape the allocation of time and resources—citizen action may level off or go along more quietly. So it is likely to be in the case of genetic engineering technology, and some day soon citizens will again take the initiative to safeguard their children's future.

The Value of Standards

Elsewhere I have argued that by protecting people from disease and injury, health and safety standards can enlarge human freedoms.[39] Traffic lights, a familiar and long-accepted form of social control, regulate when an individual can cross the street. Precisely because they contribute to *freedom from* death and injury in crossing the streets, traffic lights actually create a condition for exercising the *freedom to* cross, or drive, safely. The point is that democratic governance is nourished and sustained when the symbiotic tension between freedom and control is understood and applied. Moreover, if standards are developed in a representative and bureaucratically unencumbered manner, they can integrate the views of many affected parties and thus achieve greater legitimacy.

Mandatory government standards can serve other useful purposes. Prevention is one. The public collection of information about genetic engineering developments facilitates identification of dangerous trends or misuses *before* damage is done to health and environment. Ideally, standards are meant as preventive measures, rather than as a basis for determining legal liability after someone has been hurt. A system of standards that generates critical public information encourages the flow of technical communications and expedites public comment and criticism. In the case of recombinant DNA, standards are also useful as incentives for developing less risky sources, vectors, and hosts. But the NIH has not pursued this route fully since Roy Curtiss III and his team at the University of Alabama produced a weakened strain of *E. coli* to meet the biocontainment guidelines for what were then regarded as the most dangerous experiments.[40]

It has been suggested that molecular biologists regularly use the organisms that they understand the best in their experiments even if those organisms present more risks than other, safer organisms.[41] This practice, combined with commercialization, creates a momentum to get on with experiments at the expense of finding alternatives for safer experimentation.

Federal standards often serve as a starting point for local ordinances, which may be even stricter. "A growing number of communities," reported *Business Week*, "are making it painfully clear that they are still not comfortable living with an unregulated industry that alters the genetic structure of organism."[42] On 24 July 1981, Boston joined Berkeley and Emeryville in California, Princeton, New Jersey, and Cambridge, Waltham, and Amherst in Massachusetts by passing an ordinance covering gene-splicing companies in its jurisdiction.

Moreover, the very fact of national standards keeps the question of accountability on the front burner for all investigators and their academic and corporate leaders. In the event of an accident, who will be held accountable and to what extent? Without standards, one analyst noted, "Anybody can do anything at any place."[43]

Finally, standards serve government's essential public health responsibilities. It is, Judge David Bazelon has observed, "through the regulation of health risks that government tries to control and direct the powerful but dangerous tools of progress."[44] People look to their government to pick up the pieces after an accident, and it will be their tax money that will pay for the results of wrongdoing, or human error. The Three Mile Island nuclear accident is a pointed example of "the victims pay."[45] By supporting strategies to weaken or abolish the government's legitimate role in preventing or reducing hazards, the scientific community is setting itself up for a fierce public backlash when something does go wrong. And a biotechnology accident would be seen as much worse than a radiation accident because the public perception is that recombinant research is dealing with the very stuff of life—with the genetic inheritance. In short, scientists will be just as intimately linked with the risk as with the benefit of biotechnology.

What is the worst that can go wrong? In the case of nuclear power, Oak Ridge National Laboratory scientists and engineers were more comfortable with the benefits scenarios than with the so-called maximum scenario, a meltdown of the reactor's nuclear core with radioactive contamination of a wide area. "Probability specialists" argued that the probability of an accident was so small, why worry about consequences? And for unexpected problems there were always tech fixes. For "consequence specialists," however, probability statements were irrelevant. Their business was with the consequence of an accident and contingency planning through improved engineering design. And today, the nuclear industry is in great trouble, in part because it did not pay sufficient attention to consequence specialists and develop a first-rate quality control capability. The government's Atomic Energy Commission, too, virtually gutted its quality control programs and relied instead on computer simulations, leaving its successor organizations little in the way of this essential competence.

The point that was missed was that, though the probability of an accident may be small, the consequences could be devastating. Furthermore, as Irvin Bross, a statistician at Roswell Park Memorial Institute in Buffalo, argues, probability statements mean nothing in the face of large technologies with little operating experience. In 1976, rDNA researcher Roy Curtiss made a related point; he emphasized that probability estimates lack meaning without strict enforcement of the NIH guidelines, furthermore, he favored investigating the worst kinds of biohazards, eschewing no-risk or low-risk approaches. Nonetheless, like the nuclear power advocate, rDNA supporters reason that the probability of something going wrong is small. They seem also to believe that they can take care of any problem that crops up, just as nuclear specialists believed in their own capabilities.[46] So, more and more, certain experiments are declared safe and the strictness of the laboratory conditions under which these can be done is progressively relaxed. Forgotten are the several violations of the NIH guidelines, reminders that in deciding research priorities and practices, it may be a good idea to give consequence questions a permanent hearing.

Research on consequences aims at prevention strategies, especially pertinent as industrial scale-ups of recombinant DNA experiments proceed. In establishing a working group on the problems of scale, the NIH's Recombinant Advisory Committee has ostensibly recognized its importance. But the RAC has been careful to note that in determining the safety of the genetic material used in experiments, it was not also giving assurances about the safety of work when microorganisms are produced on a very large scale.[47] Yet, it is essential that the government set exemplary standards for safe industrial production practices. Even if not mandatory at present, these standards would convey the government's ground rules and its expectation of industry adherence. Otherwise, corporate research, development, and production will proceed without any external constraints.

Self-regulation by the biotechnology industry does not appear promising under conditions of heated races for the market, and historically, corporations have resisted government health and safety standards.[48] Moreover, union leadership is still in flux. "No union," comments Anthony Mazzochi of the Oil, Chemical, and Atomic Workers, "is deep into dealing with rDNA and that is because they don't see it. It's still in its infancy. But in this decade we're going to see [industrial] production bloom."[49] Through Mazzochi, the OCAW has been involved from the beginning. On his instruction, for example, a manual on gene splicing has been prepared for workers, recommending interim protection procedures until regulations covering the industry are in place.[50]

Experience shows that the time to bring a technology under risk management is when it is first being developed and before it becomes an uncontrollable instrument of economic power. Experience also shows the wisdom of involving the technology's toughest critics, wherever they are found, to ensure a broad expression of societal concerns.

Rights and Remedies

Governing a technology democratically depends on citizens exercising their participatory rights. These include the right to know about possible hazards as well as benefits; the right to alternative solutions, that is, less risky ways to achieve the claimed benefits; the right to secure essential information; the right to speak freely about hazards; and the right to initiate direct action.

The Right to Know about Hazards (and Benefits)

The hazard side of a technology usually takes a back seat to the benefit side. So the public has heard much more about the potential benefits of genetic engineering than its possible hazards. But as mentioned earlier, there are many uncertainties about hazards from the design of new life forms, and the prospects are disquieting. The President's Commission for the Study of Ethical Problems in Medicine and Biomedical and Behavioral Research considered a worst case scenario. In its 1982 report *Splicing Life: The Social and Ethical Issues of Genetic Engineering with Human Beings*, the commission discussed the riveting prospect of using genetic engineering to develop human-animal hybrids that may reproduce themselves, a "group of virtual slaves, partly human, partly lower animal—to do people's bidding." The report reminds us that "[d]ispassionate appraisal of the long history of gratuitous destruction and suffering that humanity has visited upon the other inhabitants of the earth indicates that *such concerns should not be dismissed as fanciful* [emphasis added.]"[51] Fourteen years earlier, James Watson, noting the rapidity of molecular cloning work, cautioned a congressional forum:

> This is a matter far too important to be left solely in the hands of scientific and medical communities. . . . A serious effort to ask the world in which direction it wishes to move [might find molecular cloning confronting] a blanket of world-wide illegality. . . . [If] we do not think about the matter now, the possibility of having a free choice will one day suddenly be gone.[52]

The process by which citizens learn about a problem often is fitful, uneven, and incremental. However, citizen action has tended to integrate at the advocacy level what is fragmented at the bureaucratic and even at the scientific levels. For example, a senior soil scientist in a government laboratory once told me that until environmentalists began to question the polluting effects of nuclear power, research on plutonium in the environment was given short shrift. Soil scientists played second fiddle to physicists and engineers who strongly backed nuclear power and focused mainly on the reactor. But as the environmental movement grew, it became more difficult to cover up plutonium spills, and the soil scientist was able to take an analytical sample and so begin the critical research on how questioning citizens have pushed research and engineering designs to higher levels of performance.

The Right to Less Risky Alternatives

Advocates of auto safety cheered the Department of Transportation's experimental vehicle program, which produced a fuel-efficient car with noteworthy safety features at a cost in 1981 of under $8000, if 300,000 of them were manufactured. Although of practical interest, few people knew about this pace-setting alternative. However, when then National Highway Traffic Safety Administrator Joan Claybrook showed the car on the Phil Donahue television show in January 1982, eight million viewers had their vision of the technological possibilities expanded. Such information, when popularly and frequently publicized, can help build a persistent constituency for buyer-oriented solutions.

Social strategies to prevent disease and accidents are as important as technologies to produce social benefits. The strategy of organizing workers as monitors of hazards on the job and as informants in the research process is one example. Information from victims has substantially helped investigators, as Dr. Irving Selikoff, who pioneered research on occupational disease at Mt. Sinai Hospital in New York City, readily acknowledges. However, basic to worker participation is understandable information about the hazards.[53]

Some have suggested that there are other ways to gain the benefits that advocates claim rDNA technology will produce. Cleaning up workplaces and the environment can prevent some cancers, for example, without the risks associated with gene-splicing. Tankers designed to prevent oil spills in the first place would obviate the need for creating microorganisms to eat oil spills and would prevent the risks that this application can entail. In short, credible alternatives enlarge people's freedom of choice.

The Right to Information

Fundamental to popular assessments are mechanisms to facilitate early access to information and ensure due process. The groundbreaking Freedom of Information Act of 1966 and the strengthening amendments of 1974, one of the victories of a mobilized citizenry, has ensured people's assess to information the federal government assembles.[54] Many examples can be cited from the food, drug, environmental, energy, political rights, and other areas, wherein citizens have used this law in support of their causes.

The early 1970s also saw the passage of the Federal Advisory Committee Act which, despite its inadequacies, formalized a role for the public members in the government's advisory apparatus and opened federal advisory committee meetings to the public.[55] Precedent-setting lawsuits and citizen petitions were also used to secure rights and reach for remedies.

Disclosure requirements are essential. Just as there are legal requirements to report pollution of water from an industrial plant, there should be similar data on the biotechnology enterprise. These should be regularly available to citizens, in disaggregated form so that problems with specific companies or universities can be resolved. But confidentiality claims or trade secrecy restrictions often obstruct the acquisition of health and safety information. When this happens with genetic engineering information, the activity can be said to be essentially outside the law. Increasing numbers of arrangements between universities and biotechnology companies should generate the most serious attention to internal corporate governance.[56]

The Right to Speak Freely about Hazards

Operating as they do on the frontlines of a technology, technical specialists can be early monitors of trouble. They must be free to speak in a public forum about harmful effects without fearing punishment, through either loss of their jobs or demotion by their superiors.[57] Unfortunately, there are disturbing examples of employers muzzling or intimidating scientists and engineers to the detriment of the public interest.[58] Such practices, as they become more commonly known, are giving rise to various remedies for protecting individuals who will put their conscience above loyalty to their organization. This is particularly important in the corporate setting because there is no union or tradition to safeguard scientists' rights.

Protecting Whistleblowers. Michigan, recognizing that the right to speak about health and safety hazards on the job should be secured, became in April 1981 the first state in the country to pass a "Whistle Blowers Protection Act." This pioneering law arose out of Michigan's PBB (polybrominated biphenyls) tragedy in the mid-1970s when a chemical company mistakenly shipped a poisonous fire retardant to state feed grain cooperatives, believing it to be their regular nutritional supplement. What spurred legislation was the testimony that employees had been warned that they would be fired if they volunteered information to investigators that the PBB accident might have caused the deaths of a large number of farm animals. "This delayed our discovering the truth," said the sponsor of the Michigan law, State Representative James Barcia, "and dealing with it."[59] Legal remedies of this kind can establish an environment of truth telling.

"Duty Arises out of the Power to Alter the Course of Events." Applying Alfred North Whitehead's dictum, what have recombinant DNA scientists done to educate people since they first expressed their concerns about biohazards only? Are they trying to broaden their public base, or are they reluctant to enlarge people's understanding because they fear a public response not to their liking?

Expressions of concern that culminated at Asilomar in 1975 were aimed more at alerting the scientific community than at educating the public. It was understandable that molecular biologists would seek to consult their colleagues about problems in the field. The dilemma in which they found themselves, where critical information was lacking but where decisions had to be made regarding how to proceed on the biohazard issue, generated confusion and controversy among the scientists. But as public concern intensified, the scientific community closed ranks. Today only a few scientists continue their public questioning of the advisability of pursuing recombination. Now and then a cautionary piece will appear in the newspapers, such as Cavalieri's 1982 opinion editorial, warning against permitting "the biosphere, surpassing as it does our understanding, to become an experimental subject. There is only one Earth," he wrote, "one earthly biosphere, and we are part of it. *There is no margin for error*."[60] (Emphasis added.)

Isolated corporate and scientific power and the growing role of universities in the biotechnology industry emphasize the need for scientists to inform the citizenry about gene-splicing as it develops. Recalling Whitehead's dictum, what should be their public education role?

Ideally, scientists would accept the public as peers in discussing normative issues relating to recombinant DNA, for on questions of values there are no specialists. On how best to achieve certain benefits, scientists would explain the advantages of this technology, as they see them. But they have a concurrent responsibility to reveal the risk side of the work also. They would explore their obligation to the communities hosting genetic engineering laboratories and companies. They would avoid jargon so that people can understand how the technology may affect them. Finally, their professional organizations would give the public education function of scientists important status and corresponding rewards so as to protect scientists against peer harassment and other punitive reactions.

Thus, enough specialists would speak out on the subject to keep people up to date about developments in genetic engineering and the issues dividing its leading advocates and critics. With public comment about their operations and intrascientific competitions, scientists as a group may come to understand that in a democracy all citizens have a duty as well as a right to assess a technology, particularly one with such far-reaching implications.

The Right to Initiate Direct Action

In his book *From Know-How to Nowhere*, Elting Morison has proposed what amounts to the establishment of community assessment boards. With the help of staff specialists, such boards would regularly hear all viewpoints, for example, on the siting of gene-splicing operations mesh with community values. Developed over a period of time in different localities, the guidelines could be viewed much as the common law is, as accepted principles to help communities decide critical issues. And, Morison argues, these principles are likely to be widely accepted because of the democratic processes that shaped them.[61]

Wisconsin's innovative Citizens Utility Board (CUB) resembles Morison's proposal for a community-based assessment and oversight capability. Utilities are required by state law to insert periodically a collection card in their billing envelope. Using this card, citizens who pay electric, gas, telephone, and water utility bills can contribute money to form and support their own CUB. The CUB would represent their interests in hearings on utility rates and on any number of other utility matters affecting health and safety as well as pocketbook issues.[62]

A full-time, citizen-run organization monitoring genetic engineering technology could be an effective instrument for critical evaluations and actions. It would have to develop diversified channels of information, especially as it becomes more difficult to find independent assessors. News media that report more consistently than at present on risks and on the arbitrary use of power could help citizen assessments, but, as an experienced science reporter remarked, science reporting is "too capricious." Many groups do publish their own reports and periodicals, some of which for over a decade have featured groundbreaking stories and raised fundamental questions.[63]

Obstacles to the Democratic Shaping of Technology

One of the most serious obstacles to a democratic process is the pressure exerted on scientists and engineers by corporations (and sometimes government agencies) to keep quiet about disturbing findings. This muzzling and intimidation are an aspect of the U.S. power structure in relation to science and technology. The mind-sets of scientists and engineers with corporate affiliations have more to do with corporate norms, stressing trade secrecy and restriction of information, that with the norms of science, stressing truth and open communication. For example, Johns Manville scientists reportedly long knew about the harmful health effects of asbestos, but did not say so publicly. Another instance involves Dow Chemical Company of Midland, Michigan. It was reported in the *Washington Post* that, according to a 1965 internal memorandum, Dow scientists were concerned about health damage from dioxin and reported to other chemical firms that dioxin caused "severe" liver damage in rabbits, but this information was not given to the people of Midland for fear of a sharp public reaction and more government regulations. "Group-think" is how one scientist from the Michigan Department of Natural Resources describes the behavior of the Dow Scientists. A leading citizen activist in the community lamented that "the scientists at Dow are a valuable resource, but open discussion is not encouraged among them."[64]

The early debate on rDNA ended with responsibilities for both development of the technology and its public health implications centralized in the director's office of the National Institutes of Health, the agency sponsoring the research. The Occupational Health and Safety Administration, the National Institute of Occupational Health and Safety, and the Environmental Protection Agency might have played a role, but did not. The conflict of interest so created has not been overcome. When the ability to create and promote a technology is confused with the ability to control it, a powerful bias for the technology is created, as indeed we now see with genetic engineering. The idea that sponsors of a technology can regulate it persists, despite experience that when the same government agency is responsible for development *and* public health, as with the Atomic Energy Commission, facts seem not to matter as much as investments, egos, and careers.

A second obstacle to the democratic shaping of technology is the power imbalance and the myth that scientific objectivity always prevails and everyone can have their say. Promoters of a technology can be expected to be unsympathetic to assessments, no matter how well documented, that question their definitions of benefits and hazards. The promoters dominate the course of the technology's development and its information flow, and the critics too frequently are without a formal voice in the decision-making structure.

The recombinant technology suffers from this power imbalance. The orientation of the NIH advisory committee was toward promotion of the technology, and the original guidelines and oversight procedures have been steadily dismantled. Minimal sanctions for violators and the absence of NIH leadership in conducting exchanges between eminent scientific critics and promoters to clarify the technical issues for the public have compounded the problem.

Finally, the "urgency factor" militates against moving carefully to insure the general welfare. Recombinant enthusiasts claim that the work must proceed at once because the potential benefits for human health, for food production, and so on, are so great. But should that be the criterion? As Daniel Singer said at the Asilomar biohazard conference, "Big benefits do not justify big risks."[65] The harmful effects of this fundamentally antidemocratic structure for technology development have in the past been expressed in very human terms—birth defects, hazardous drinking water, premature deaths, and other kinds of workplace and community injuries and diseases, and in some cases ever higher risk levels.

The constituency for risk management may not be large at present, but that does not mean that risk management is either unimportant or inappropriate. Gene-splicing specialists have a special responsibility to assess the risks to the public associated with their particular science and technology. As one Cambridge city councillor asked of a Harvard scientist, "Just what . . . do you think you're gonna do if you *do* produce [a dangerous organism]?"[66] There was no answer from the scientist. To the question "What is the worst scenario?" there is again a virtual silence on the part of rDNA advocates. But this is the fundamental question that must be asked along with any consideration of benefits. The public has a right *and* a duty to ask about hazards. So do the scientists and engineers.

Unfortunately, a commitment to continuous risk assessment work is hampered by an attitude which David Baltimore, a central figure in the biotechnology enterprise, aptly summarized: "In the end, you know, cosmic questions are probably an old man's game."[67] This perspective, encouraged by the race for breakthroughs in biotechnology and the expected economic rewards, tends to push aside critical questions of responsibility, accountability, and ethical behavior. It hinders essential public learning, a process which should involve the gene-splicing scientists and engineers. For, as Liebe Cavalieri reminds us,

> Genetic engineering is not just another scientific accomplishment. Like nuclear physics, it confers on human beings a power for which the [molecular biologists] are psychologically and morally unprepared. The physicists have already learned this, to their dismay; the biologists, not yet. "Indeed, one Nobel laureate has boasted, *'We can outdo evolution.'*"[68] (Emphasis added.)

Such assertions highlight the need for thorough investigations of associated risks. Yet, by connecting themselves directly with its commercial development, those who created biotechnology are, therefore, inappropriate candidates for such critical analyses. A growing number of citizens must join the discussions about gene-splicing and its repercussions if biotechnology companies and researchers are not to escape democratic oversight. A concerned public must become well informed if it is to develop an organized capability for ensuring full and independent assessments of the hazards. For, as James Madison observed, "a people who mean to be their own governors must arm themselves with the power knowledge gives."[69] In such matters as this, an important power is the power of review. The principle of democratic oversight is basic to our society, and its application is essential for producing popular information and popular decisions. The social consequences of the recombinant DNA technology are too enormous and important to be left to specialists alone.

Notes

1. Liebe F. Cavalieri, "Genetic Engineering: A Blind Plunge," *Washington Post,* 14 May 1982, A31.
2. See, for example, Mary Kaldor, "Technology and the Arms Race," *The Nation,* 9 April 1983, pp. 420–23. The entire issue covers different aspects of the arms race.
3. Some specialists also did not understand the probabilities based on their information or lack of it. See, for example, Daniel Ford, *The Cult of the Atom: The Secret Papers of the Atomic Energy Commission* (New York: Simon & Schuster, 1982), pp. 67–74.

4. For a history and analysis of the contemporary citizen movement in the United States, see Harry C. Boyte, *The Backyard Revolution: Understanding the New Citizen Movement* (Philadelphia: Temple University Press, 1980). See also the collection of essays by community organizer Byron Kennard, *Nothing Can Be Done, Everything Is Possible* (Andover, Mass.: Brick House, 1982).
5. MIT biologist Jonathan King argues that taxpayers also should realize monetary benefits that may flow from recombinant DNA work. See his testimony, "Protection of Publicly Available University Biomedical Research Is Essential to the Nation's Health," in *Hearings,* Committee on Science and Technology of Subcommittee in Investigations and Oversight, U.S. Congress, House of Representatives, 8 June 1981. Philip Siekevitz of the Rockefeller University makes the same point in a letter to the editor of *Science* 219 (4 March 1983): 1022.
6. For an account by a leading nuclear safety advocate, see Ford, *Cult of the Atom.*
7. For specific U.S. and foreign examples of how attractive energy efficiency and renewable energy options are becoming, see Amory Lovins, *Brittle Power* (Andover, Mass.: Brick House, 1982), pp. 223–31.
8. Quoted in John Lear, *Recombinant DNA: The Untold Story* (New York: Crown, 1978), 159.
9. Quoted in Liebe F. Cavalieri, "Twin Perils: Nuclear Science and Genetic Engineering," *Bulletin of the Atomic Scientist* 38 (December 1982): 73.
10. J. Kenneth Galbraith, *The Affluent Society* (Cambridge, Mass.: Riverside, 1958).
11. For a comprehensive history of the American populist movement, see Lawrence Goodwyn, *Democratic Promise: The Populist Movement in America* (New York: Oxford University Press, 1976).
12. For example, see Ralph Nader, Peter J. Petkas, and Kate Blackwell, eds., *Whistleblowing* (New York: Grossman, 1972). See also Constance Holden, "Scientist with Unpopular Data Loses Job," *Science* 210 (14 November 1980): 749–50.
13. For illustrations of this point with regard to other technologies, see Barry Jones, *Sleepers Wake! Technology and the Future of Work* (Melbourne: Oxford University Press, 1982), pp. 231–32.
14. Quoted in Cavalieri, "Genetic Engineering."
15. For a discussion of risks, see Chapter 4 in Richard Hutton, *Bio-Revolution: DNA and the Ethics of Man-Made Life* (New York: New American Library, 1978).
16. See Meredith Tushen, "Why OCAW Members Should Know about Recombinant DNA Research," booklet prepared by the Health and Safety Department, Oil, Chemical, and Atomic Workers International Union AFL-CIO, p. 8, for a recommended set of procedures for interim protection pending regulations covering industry, or stopping rDNA work; Ditta Bartels, "Occupational Hazards in Oncogene Research," *Gene Watch* 1 (September–December 1984): 6–8; for greater detail, see note 33.
17. On 14 September 1983, a group of environmentalists filed suit against the National Institutes of Health on grounds that the agency did not comply with the National Environmental Policy Act when it approved the first release of geneti-

cally altered bacteria into the open environment before assessing how the action could change the ecology of an area. The proposed experiment was aimed at replacing bacteria contributing to frost on plants with genetically changed bacteria that would not produce frost. See *New York Times,* 15 September 1983, A17.

18. The letter from Paul Berg et. al. (who constituted The NAS Committee on Recombinant DNA Molecules, Assembly of Life Sciences), published in *Science* and *Nature,* called for a worldwide moratorium by scientists on two kinds of experiments considered potentially hazardous pending a review of the hazards. See Judith P. Swazey, James R. Sorenson, and Cynthia B. Wong, "Risks and Benefits, Rights and Responsibilities: A History of the Recombinant DNA Research Controversy," *Southern California Law Review* 51 (September 1978): 1023–26 for the full substance of the Committee's recommendations. For accounts of the rDNA controversy see other chapters in this volume; see also Nicholas Wade, *The Ultimate Experiment: Man-Made Evolution* (New York: Walker, 1977); Lear, *Recombinant DNA*; and Hutton, *Bio-Revolution.*
19. See Nicholas Wade, *The Ultimate Experiment,* Chapter 11, for a brief summary of these events.
20. Quoted in Wade, *The Ultimate Experiment,* p. 131.
21. Susan Wright, "Molecular Politics in Great Britain and the United States: The Development of Policy for Recombinant DNA Technology," *Southern California Law Review* 51 (September 1978): 1383–1434.
22. Susan Wright, "The Status of Hazards and Controls," *Environment* 24 (July/August 1982): 12ff. See also Stuart A. Newman, "The 'Scientific' Selling of rDNA," *Environment* 24 (July/August): 21ff.
23. In a letter to Dr. William Gartland of the Office of Recombinant DNA at the NIH, Francine Simring, executive director of the Coalition for Responsible Genetic Research, writes, "Recent broad exemptions and elimination of prohibitions in recommended DNA research by the NIH make it especially impelling to maintain the guidelines as regulations." (29 July 1981). On the Supreme Court decision see, Sheldon Krimsky, *Genetic Alchemy: The Social History of the Recombinant DNA Controversy* (Cambridge, Mass.: The MIT Press, 1982), pp. 204, 292, 347.
24. William Bennett and Joel Gurin, "Science That Frightens Scientists: The Great Debate over DNA," *Atlantic Monthly* 239 (February 1977): 44.
25. Katherine Bouton, "Academic Research and Big Business: A Delicate Balance," *New York Times Magazine,* 11 September 1983, pp. 62ff.
26. Quoted in Joann Rodgers, "Asilomar Revisited," *Mosaic* 12 (January/February 1981): 25. See also *New York Times,* 22 May 1983, p. 29, on Harvard University's guidelines for faculty involved in corporate-sponsored research.
27. John T. Edsall, "Two Aspects of Scientific Responsibility," *Science* 212 (3 April 1981): 11–12.
28. David F. Noble, "The Selling of the University," *The Nation,* 25 February 1981, pp. 129ff.
29. For a review of this process, see Wright, "Status of Hazards and Controls," pp. 16ff.
30. See Diana Dutton and John L. Hochheimer, "Institutional Biosafety Commit-

tee and Public Participation: Assessing an Experiment," *Nature* 297 (6 May 1982): 11–15; Daniel F. Liberman, "Biosafety Committees: Protection in the Lab and Community," *Environment* 24 (July/August 1982): 14–15; and Phillip L. Bereano, "Institutional Biosafety Committees and the Inadequacies of Risk Regulation," *Science, Technology, and Human Values* (Fall 1984): 16–34.

31. Wright, "Status of Hazards and Controls," pp. 20, 51.
32. Philip J. Hilts, "NIH Unit Would Unleash Gene-Splice Experiments," *Washington Post,* 11 September 1981, p. A2.
33. A 1979 report to the Commission of European Communities expressed concern about possible hazards of this kind emanating from the industrial use of microorganisms. See Wright, "Status of Hazards and Controls," p. 16. See also Ditta Bartels, Hiroto Naora, and Atuhiro Sigatani, "Oncogenes, Processed Genes, and Safety of Genetic Manipulation," *Trends in Biochemical Sciences* 8, no. 3 (March 1983): 78–80; Ditta Bartels, "Oncogenes: Implications for the Safety of Recombinant DNA Work," *Search* 14, no. 3-4 (April/May 1983): 88–92.
34. *The Environmental Implications of Genetic Engineering,* Hearings held by the Subcommittee on Investigations and Oversight, House Committee on Science and Technology, 22 June 1983. A staff report of same title was issued in February 1984. See also *Gene Watch* 1 (September–December 1984); 10–13 for a brief review of the subcommittee's work on biotechnology in the early 1980s.
35. Kim McDonald, "EPA Seeks Role in Approving Genetic Tests," *The Chronicle of Higher Education* 27 (December 14, 1983): 1, 26.
36. See note 17.
37. Al Wyss, "$10 Billion Market Forecast for DNA," *Journal of Commerce,* 20 April 1981, p. 1; Peter Behr, "Boom or Bust in the Biotech Industry," *Environment* 24 (July/August 1982): 6–7; Dick Russell, "The Marketing of Genetic Science: The Tree of Knowledge Grows on Wall Street," *The Amicus Journal* 5 (Summer 1983): 14–23; "Biotech Comes of Age," *Business Week,* 23 January 1984, pp. 84–94.
38. To help such communities, a model ordinance was developed. See David Ozonoff, "A Model Ordinance: Protection at the Local Level," *Environment* (July/August 1982): 18–19; also Sheldon Krimsky, Anne Baeck, and John Bolduc, *Municipal and State Recombinant DNA Laws: History and Assessment* (June 1982), report available from Krimsky, Department of Urban and Environmental Policy, Tufts University, Medford, Mass. 02155; see also Chapter 22 on "Local Initiations for Regulations" in Krimsky, *Genetic Alchemy.*
39. Claire Nader, "Controlling Environmental Health Hazards: Corporate Power, Individual Freedom, and Social Control," in *Public Control of Environmental Health Hazards, Annals of the New York Academy of Science* 329 (26 October 1979): 213–20.
40. See discussion in Wade, *The Ultimate Experiment,* pp. 67–84.
41. On this point see Roger G. Noll and Paul A. Thomas, "The Economic Implications of Regulation by Expertise: The Guidelines for Recombinant Research," in *Research with Recombinant DNA,* proceedings of the Academy Forum (Washington, D.C.: National Academy of Sciences, 1977), pp. 266–68.

42. *Business Week* (industrial edition), 10 April 1981, p. 32; see also Halsted R. Holman and Diana B. Dutton, "A Case for Public Participation in Science Policy Formation and Practice," *Southern California Law Review* 51 (September 1978): 1505–54.

43. W.A. Thomasson, "Recombinant DNA and Regulating Uncertainty," *Bulletin of the Atomic Scientists* 35 (December 1979): 31.

44. "Bazelon Calls for Disclosure of Health Risks," *Health Sciences Report,* Harvard School of Public Health, Summer 1981, p. 1.

45. In a move to socialize the costs of the Three Mile Island accident, the Reagan administration announced in October 1981 that it would help pay for the cost of cleanup.

46. See Cavalieri, "Twin Perils," 72–75, for a provocative comparison of today's molecular biologist with the nuclear power enthusiasts in the first flush of their accomplishments; for Roy Curtiss' position, see Swazey et. al., "Risks and Benefits, Rights and Responsibilities: A History of the Recombinant DNA Research Controversy," pp. 1050–51; Bross' view was expressed at a public meeting in the 1970s in Washington, D.C., which I attended.

47. Kathy Yih, "Biotechnology Becomes Big Business," *Science for the People* 12 (September/October 1980): 7.

48. See Mark Green and Norman Waitzman, *Business War on the Law: An Analysis of the Benefits of Health/Safety/Enforcement,* rev. 2nd ed. (Washington, D.C.: Corporate Accountability Group, 1981).

49. Telephone communication, 25 May 1983.

50. See also Tushen, "Why OCAW Members Should Know about Recombinant DNA Research."

51. President's Commission for the Study of Ethical Problems in Medicine and Biomedical and Behavioral Research, *Splicing Life: The Social and Ethical Issues of Genetic Engineering with Human Beings* (Washington, D.C.: GPO, 1982), p. 58; see also the section entitled "Concerns About 'Playing God,'" pp. 53–60. The following year at a press conference on 8 June 1983, leaders of the U.S. religious community called upon Congress to prohibit the engineering of specific genetic traits into the human germ line. See *New York Times,* 9 June 1983, A2.

52. Quoted in Lear, *Recombinant DNA,* p. 159. One of the purposes of the Committee for Responsible Genetics (Cambridge, Mass.), successor to the Committee for Responsible Genetic Research, is to strengthen the public's role in setting policy for biotechnology.

53. A start was made by the New York Academy of Sciences when it published for workers the major findings of the first international meeting on occupational cancers. See Phyllis Lehmann, *Cancer and the Worker* (New York: New York Academy of Sciences, 1977).

54. See Francesca Lyman, "Locking up Federal Files," *Environmental Action* 13 (July/August 1981): 22 and 24, for a review of the status of the Act and the Reagan administration's attempt to narrow the law and reduce citizen access to

information from the federal government. This effort is continuous as is the opposition from citizen groups.

55. Kit Gage and Samuel S. Epstein, "The Federal Advisory Committee System: An Assessment," *Environmental Law Reporter* 7 (February 1977): 50001–12.
56. For a detailed discussion of corporate governance issues, see Ralph Nader, Mark Green, and Joel Seligman, *Taming the Giant Corporations* (New York: Norton, 1976).
57. See, for example, Rosemary Chalk and Frank von Hippel, "Due Process for Dissenting 'Whistle-Blowers'," *Technology Review,* June/July 1979, pp. 49–55. See also Rosemary Chalk, Mark S. Frankel, and Sallie B. Chager, *Professional Ethics Activities in the Scientific and Engineering Societies,* a publication of the Professional Ethics Project of the American Association for the Advancement of Science (Washington, D.C.: AAAS, 1980).
58. For an example of a scientist who fought back, see Marjorie Sun, "A Firing over Formaldehyde," *Science* 213 (7 August 1981): 630–31.
59. Alan F. Westin, "Michigan's Law to Protect the Whistle Blowers," *Wall Street Journal,* 13 April 1981, p. 18.
60. Cavalieri, "Genetic Engineering."
61. Elting E. Morison, *From Know-How to Nowhere: The Development of American Technology* (New York: Basic Books, 1974), pp. 162–87.
62. See State of Wisconsin, 1979 Assembly Bill 2, Chapter 72, Laws of 1979 for "The Citizens Utility Board Act."
63. For a representative sample of the periodical literature, see *Periodicals of Public Interest Organizations: A Citizen's Guide,* compiled and published by Commission for the Advancement of Public Interest Organizations (Washington, D.C., 1979, updated Fall 1982), pp. ix–57.
64. Ward Sinclair, "Dioxin Brings Dow under Fire." *Washington Post,* 24 April 1983, p. A1; for a more complete account, see Keenen Peck, "A Company Town Makes Peace with Poison," *The Progressive* 47 (June 1983): 26–29; and Jeremy Main, "Dow versus the Dioxin Monster," *Fortune* 107 (30 May 1983): 83ff.
65. Quoted in Lear, *Recombinant DNA,* p. 139.
66. Quoted in Bennett and Gurin, "Science That Frightens Scientists," p. 60.
67. Quoted in Rodgers, "Asilomar Revisited," p. 25. According to the account of Charles A. Beard and Mary R. Beard, biologist Edwin Grant Conklin, in 1937, almost fifty years ago, as president of the American Association for the Advancement of Science, asserted that both science and ethics are necessarily part of programs for human welfare; he advocated their unity in the thought and action of scientists. Conklin reminded his associates that "free thought, free speech and free criticism are the life of science." Moreover, "In spite of a few notable exceptions," he pointed out, "scientists did not win the freedom which they have generally enjoyed, and they have not been conspicuous in defending this freedom when it has been threatened." See their work *American in Midpassage,* Vol. II (New York: The Macmillan Company, 1939), pp. 850–51.

68. Cavalieri, "Genetic Engineering." The recent scientific findings that a nuclear war would result in a "nuclear winter, where plants, animals, and people cannot survive" give his words a special poignancy. See, for example, R. P. Turco, O. C. Toon, T. P. Ackerman, J. B. Pollack, and Carl Sagan, "Nuclear Winter: Global Consequences of Multiple Nuclear Explosions," *Science* 222 (23 December 1983): 1283–92.

69. In a letter to W. T. Barry on August 4, 1822.

From the earliest days of the rDNA controversy, there were those who voiced concern about the possibility of employing genetic engineering techniques for illicit purposes; that is, to perpetrate terrorism or wage war. In view of the extraordinary sanguine promises that the revolutionary techniques of molecular biology appear to hold for the betterment of mankind, is there also a promise for the reverse? Is it churlish to suggest that there may exist nefariously inclined decision-makers, who when deprived of the possibility of wielding nuclear arms will look toward other instruments of mass destruction? As recent events in the Arabian Gulf region demonstate, the notion is not so farfetched, nor should it be ignored. This being so, Zilinskas examines the potential that genetic engineering holds for the darker side of human endeavors and the mechanisms that are in place which attempt to surpress the actual expression of that side.

His view is that of a clinical microbiologist who after many years of day-to-day association with virulent pathogens has gained an appreciation of the fine balance existing between man (the potential host), the parasites that prey on him, and the environment shared by both. Biological weapons may destructively affect two facets of this balance—parasite and the environment. In the future, biotechnology could give rise to substances that would attack man; though this aspect of biological warfare is not dealt with in detail here. Yet the question remains, does genetic engineering offer possibilities by merely adding destructive force, or will it occasion an entirely new situation? If so, will it be necessary for us to rethink the basic underpinnings of international law that seek to prevent biological warfare?

Zilinskas believes there is still time to consider calmly and rationally the potential the new techniques of biotechnology hold for biological warfare. In particular, some hard thinking has to be done before the next review conference of the 1972 Biological Warfare Convention is held; now scheduled for 1986. After 1986 the accelerating developments in the biosciences may make so much headway that little can be done to arrest nasty doings.

It must be made clear that the views and opinions presented in this chapter are Dr. Zilinskas' own and do not necessarily represent the Office of Technology Assessment or any other organizations with which he has been, or is, associated with.

10

Recombinant DNA Research and Biological Warfare

RAYMOND A. ZILINSKAS

Introduction

Since the earliest days of the recombinant DNA controversy, the possibilities of applying the new technique for the design and manufacture of bacteriological (biological) weapons (BW) has been considered. The topic was briefly discussed at Asilomar in 1975 and more extensively at the National Academy of Sciences Academy Forum in 1977. In addition, various authors have written their personal views on the subject, and some of those will be cited here. The concern is international; for example, Soviet writers N. V. Turbin and O. Baryoan have stated their misgiving of Western powers using rDNA methods to create "new, death-dealing weapons."[1]

Recently, there has been an upsurge of interest in the subject because of claims made against both the United States and the Soviet Union, accusing them of illegal manufacture and use of BW. Premier Fidel Castro has held the United States responsible for initiating human epidemics of acute hemorrhagic conjunctivitis (AHC) and dengue fever in Cuba and also for various epidemics striking animals and plants. Events surrounding an outbreak of anthrax in Sverdlovsk, USSR, have raised questions regarding the possibility of an active BW facility being sited there, and more recently, the USSR has been accused of having supplied mycotoxins for use in the Indochina peninsula and, perhaps, Afghanistan.[2]

The possibility of the Soviet Union supplying, and abetting the use of, mycotoxins in warfare has apparently caused the Reagan administration to begin a reevaluation of the defensive status of the United States vis-à-vis BW. The effort expended by the United States on BW has been of low order and has been limited to defensive work since 1969, when President Nixon ordered all offensive research, development, and production unconditionally terminated. President Nixon's order was formalized when the United States adopted the 1972 Convention on the Prohibition of the Development, Production, and Stockpiling of Bacteriological (Biological) and Toxin Weapons and on their Destruction (hereafter cited as BWC).[3] Thus, the United States will not conduct offensively oriented research and development (R&D) but may increase defensive R&D involving detection, protection, and therapy. Since rDNA offers possibilities in these areas, one should assume that rDNA research and its techniques will be assessed as to their potential in BW. It is therefore appropriate to discuss recent advances in rDNA research and the opportunities they may offer to BW research, development, deployment, and employment. Once this issue has been examined, one can evaluate existing law, which seeks to prevent all BW, in light of recent advances. If it is found wanting, suggestions for palliative measures can be made. At the very least, perhaps a larger segment of the public will become acquainted with an emerging problem area and will thus be in a position to oppose the realization of yet another weapons system.

BW weapons include those disabling or lethal agents made up of living organisms or their toxic products. Suitable organisms could be bacteria, mycoplasmas, rickettsiae, viruses, fungi, or other still uncharacterized organisms. Toxins are chemical agents, conceivably useful as weapons, that include bacterial endo- and exotoxins. Though some of the most toxic substances in existence are of biological origin, none has seen more than the most limited use as a weapon. Likewise, comparatively few microorganisms have been used for warfare purposes, and the number is unlikely to increase for reasons soon to be discussed. Thus the number of organisms to be considered in this chapter is rather small—perhaps ten to twenty bacterial species and fewer than ten viruses, fungi (yeasts), and rickettsiae.

In the following sections, I will first give a short history of BW, then an account of the successful use of BW in an interspecies conflict. As a counterpart, I will present the practical restrictions that have so far prevented the employment of BW in human conflicts and then I will describe possibilities that rDNA research offers to the development of biological warfare agents. Next I will consider instruments that the international system has structured to prevent BW and point out some of their deficiencies. I will also comment briefly on the problem of terrorism.

History of Biological Warfare

No doubt, disease has decisively affected the course of mankind's history in several instances.[4] The use of disease as a weapon or as a means of war has been described infrequently in history, however, so applicable instances stand out. One instance took place in the fourteenth century after the pioneering journey of Marco Polo had led to the establishment of caravan trails by which a rich and varied trade was carried on between Asia and the important city-states of Genoa and Venice. The two cities fought incessantly, but eventually Genoa prevailed and was able to control the end point of the caravan trail—the city of Caffa on the Crimea. The Genoese fortified Caffa to the extent that by 1344, when war broke out between Genoa and the Mongols, it was considered impregnable.[5] Coincidentally, a dreadful plague broke out in the Far East (approximately 1344 A.D.), and by 1346 it had spread to the Crimea where the Golden Horde may have suffered as many as 85,000 dead. Inside Caffa all was well until:

> The Tartars, fatigued by such a plague and pestiferous disease, stupefied and amazed, observing themselves dying without hope of health, ordered cadavers placed in their hurling machines and thrown into the city of Caffa, so that by means of these intolerable passengers the defenders died widely. Thus there were projected mountains of dead, nor could the Christians hide or flee, or be freed from such a disaster . . . they allowed the dead to be consigned to the waves. And soon all the air was infected and the water poisoned, corrupt and putrified, and such a great odor increased . . . so great and so much was the general mortality that great shouts and clamor arose from the Chinese, Indians, Persians, Nubians, Ethiopians, Egyptians, Arabs, Saracens, Greeks, who cried and wept, and suspected the extreme judgement of God.[6]

Soon the Genoese quit the city, sailing from the Caffa port, which the Mongols had not been able to close. Unfortunately, the evacuees brought the plague with them, spreading it wherever they landed. DeMussis continued his remarkable account:

> . . . and we Genoese and Venetian travelers, of whom scarcely 10 survived out of 1,000, as though accompanied by a malign spirit, entered our homes . . . relatives and friends and neighbors came to us from all sides . . . woe to us who carried the darts of death, as they held us with embraces and kisses while we spoke, from our mouths we were compelled to put out poison with our words. Thus, returning to their own homes they soon poisoned the entire family.[7]

It is clear from DeMussis's account that though the medieval people knew nothing about the causes of disease, they were aware of the concept of infectivity and had an empirical awareness of disease transmission. We can also glean from his account several of the limitations of BW, including unpredictability and uncontrollability.

BW Use During World Wars I and II

During the early days of U.S. involvement in World War I, a successful operation to interfere with the shipping of livestock to the Allies may have taken place, although accounts differ as to what actually happened. Apparently, a German saboteur intercepted some animals destined for Europe and infected them with a debilitating disease. One authority[8] credited the saboteur as having injected 4500 donkeys with *Malleomyces mallei*.[9] However, G. W. Merck, a special consultant on BW to the Secretary of War, wrote in a 1946 report: "There is incontroversial evidence, for example, that in 1915 German agents inoculated horses and cattle leaving United States ports for shipment to the allies with disease-producing bacteria."[10] Unfortunately, the "incontroversial evidence" was not produced—but perhaps it was part of a classified section of the report.

During the interwar period there are no known instances of BW use, though the Japanese Army began a research effort in 1936, or earlier, to produce offensive BW weapons, an effort that continued until VJ Day in 1945.[11] According to testimony offered by eight Japanese Army officers and three enlisted men accused of directing offensive BW in Manchuria and China, research was begun in Japan as early as 1931 to prepare offensive BW devices, and in 1936, the major effort was transferred to Manchuria. Active research and development continued until the USSR's armed forces occupied the area in 1945. As part of their BW program the Japanese at first used livestock in field trials, but after 1941 they expanded their scope and began to employ captured Chinese, Russians, and perhaps Americans as guinea pigs in some rather grisly experiments. One defendant, Kawashima Kiyoshi, the former head of Detachment 731 (the unit responsible for BW developmental work), claimed that over 3000 persons had been killed in the BW experiments during 1940–45.[12] The experimental procedures were gruesome; in one instance described, ten persons were tied to stakes ten meters apart from one another. The experimental subjects were completely protected except for the buttocks. A bomb contaminated by gangrene bacteria was then exploded nearby causing wounds upon all subjects. The ten persons were untied and taken to a facility for observation where "within a week they all died in severe torment."[13]

Though the Japanese investigated several types of disease agents, including those responsible for anthrax, cholera, glanders, paratyphoid, and typhoid, the testimony indicates that plague-carrying fleas were thought to be the most effective system. The defendants all admitted to having employed BW weapons against population centers, but seemed unable to give data on results obtained. Two instances of BW employment were repeatedly described. In 1939, dysentery, paratyphoid, and typhoid bacteria were used to contaminate the waters of the river Khalkin-Gol with the hope of causing enteric disease among Soviet and Mongolian troops camped on or near the river. In the second instance, in 1940, plague-carrying fleas were scattered over Chinese troops in Central China.[14] In neither case, nor in any other case of alleged BW employment, were results known.

Later, the Chinese were to provide additional information. During the Korean War, while trying to build a case against the United States (see below), they charged the Japanese with two instances of BW use.[15] The first such effort took place in 1941 when eleven Chinese cities were attacked by Japanese airplanes that dropped bombs allegedly containing plague-harboring fleas. The second instance occurred in 1945 when cholera and plague were spread along the Chekiang-Kiangsi railroad in Central China. The Chinese said "the losses suffered by our people were inestimable."[16] Yet, curiously, the Chinese claimed only 700 victims as a result of *all* Japanese BW aggressions during 1940–44[17]—a surprisingly small number when considering that war-torn China had a large exposed population, was weakened by a devastating war, and had been attacked by BW devices containing extremely virulent pathogens.

An interesting account of the contemplated use of anthrax by the United Kingdom during the waning days of World War II was broadcast on BBC-TV's program, "Newsnight."[18] According to the program, Winston Churchill considered the use of anthrax, mustard gas, and phosgene as means of retaliation for German V-1 and V-2 attacks. He contemplated the employment of four-pound bombs developed by the United States. Each bomb was filled with 90 percent water and 10 percent anthrax bacilli. It was claimed that the United States, in 1945, was capable of producing 500,000 such bombs per month at a plant in Vigo, Indiana. However, the bombs were "grossly inefficient"—a full attack on six German cities would have required 4,117,100 bombs to kill 50 percent of the total population. Why the plan was not followed up on is unknown; most likely technical problems prevented any but the most cursory thought on the subject.

BW Offenses Since the World Wars

During the course of several wars following World War II, accusations of BW have been issued. One notable example was a series of charges made by China accusing the United States of having used BW against North Korea and China. According to the Chinese the major BW offensive took place from 19 February to 21 March 1952, when United States warplanes invaded Chinese air space and "disseminated insects and other objects" in the provinces of Jehol, Liaoksi, Liaotung, Kirin, Sungiang, and Heilungkiang.[19] Supporting evidence documenting this offensive, as well as other BW incidents, was presented before the Executive Committee of the World Peace Council in Oslo on 19 March 1953, and the Committee was asked to form an International Scientific Commission (ISC) to investigate the Chinese charges. The Chinese took this approach because it considered other international agencies, such as the International Red Cross and the World Health Organization (WHO), "not . . . sufficiently free from political influence to be capable of instituting an unbiased inquiry in the field."[20] The Executive Committee unanimously agreed to accede to the Chinese request and invitations to join the ISC were sent out to numerous scientists. The ISC was eventually constituted by six scientists and four support personnel. Members of ISC proceeded to China and met with the Academia Sinica and the Chinese Peace Committee on 28 June 1952. Shortly afterward the Commission visited stricken areas and interviewed witnesses and captured US airmen. The conclusion of the ISC was:

> The people of Korea and China have indeed been the objects of bacteriological weapons. These have been employed by units of the USA armed forces, using a great variety of different methods for the purpose, some of which seem to be developments of those applied by the Japanese Army during the Second World War.[21]

Specifically, the Commission found that the United States (1) had spread plague in China and Korea by using infected fleas; (2) had employed *Pasteurella multocida* against domestic stocks; (3) had spread the causative bacteria for cholera, typhoid fever, and dysentery through populated areas; and (4) probably had disseminated through aerosols the virus causing encephalitis over several cities where epidemics had occurred.[22]

I have discussed in detail elsewhere why I do not think the Chinese allegations were substantiated.[23] My conclusion was based on the suspicious makeup of the ISC; questions regarding the manufacture of the supposed BW delivery vehicles; shortcomings in the entomological, phytopathological, and epidemiological data presented by the Chinese; the unreliability of testimony offered by witnesses and captured U.S. airmen; and the lack of correlation between the BW agents that the United States was accused of having employed and the agents that the U.S. BW establishment had actually developed and stockpiled. In any case, the entire episode of alleged use of BW during the Korean conflict awaits a judicious historical treatment.

Possible BW Use by the Superpowers

In the following pages, two events will be discussed in which the superpowers have been accused of the illegal use of BW.

The first involved an outbreak of epidemic anthrax in the USSR. The causative organism of anthrax, *Bacillus anthracis*, is endemic to soils worldwide and has, in the past, been considered a potential BW agent. Anthrax is a disease of animals transmissable to man, sometimes through the consumption of contaminated meat. Three types of human anthrax are recognized: cutaneous anthrax, intestinal anthrax, and inhalation anthrax. The cutaneous form most frequently presents a localized lesion, self-limiting and of short duration. The other two forms usually become systemic by developing into septicemia. In this case mortality rates may range from 60 to 100 percent. The disease can be prevented by vaccination and the causative organism is susceptible to common antibiotics. However, when confronting the systemic forms, antibiotic intervention must begin quickly due to the highly virulent character of the disease. Even then there is too little published evidence to allow a reliable estimate of the effectiveness of antibiotic therapy on systemic anthrax.

The human forms of anthrax are rare in the developed world with the possible exception of the USSR. In the USSR the disease is considered to be endemic to large agricultural areas, including parts of Siberia. Though quite a good deal is known about the anthrax problem in the USSR, mystery surrounds a recent outbreak.

That outbreak took place during April 1979 in and around the city of Sverdlovsk, which is located approximately 1200 kilometers east of Moscow in the Ural Mountains. The city is remembered in history as the place where the last Romanovs met their end. At present it is a thriving industrial center populated by 1.2 million people. In addition to industry, the city is part of an important agricultural district that includes an animal population of approximately 850,000 cattle, 600,000 pigs, and 250,000 sheep and goats.[24] The city is closed to foreigners, ostensibly because of its extensive armament-related industry, perhaps including a suspected BW establishment.

The anthrax epidemic was kept secret by the USSR; only much later did Western communications media become aware of the event. The first accounts of the episode appeared in a succession of Russian emigré magazines and English and German sensationalist magazines. However, on 13 February 1980, the story was carried by the widely circulated German magazine *Bild Zeitung*, and major newspapers in the Western world began to take an interest in Sverdlovsk. Concerned that the epidemic might have been caused by the accidental release of a BW agent, the U.S. Department of State (DOS) asked T. J. Watson, Jr., U.S. Ambassador to Moscow, to inquire about the Sverdlovsk incident. The inquiry was delivered to the USSR about 16 March. On 19 March the Soviet news agency TASS denounced American suggestions as slanderous propaganda. At the same time, the Soviet government replied seriously to the State Department inquiry, acknowledging the outbreak of anthrax. However, the epidemic was blamed on bad food handling that allowed contaminated meat to be dispensed to the population of Sverdlovsk.

The potential serious consequences should not be underestimated—even now. The United States and the USSR are both signatories of the BWC. The convention obligates signatory nations to destroy their BW stocks, as well as to fulfill other convention provisions, within nine months after signing. By the end of 1975, the USSR had certified its compliance. The implications of the Sverdlovsk ''accident'' are clear. If the Soviets are still stockpiling or producing BW agents in quantities or of types not justifiable for peaceful purposes, they are in violation of international law.[25]

The second incidence of alleged BW use by superpowers may turn out to be more serious and could further damage already poor U.S.-Soviet relations. On 13 September 1981, while visiting West Berlin, then Secretary of State A. M. Haig, Jr., announced that the United States had physical evidence which "proved" the Vietnamese had used lethal biological weapons in the Indochina peninsula.[26] According to the DOS, analysis of leaf and stem samples from Kampuchea were positive for three mycotoxins—Nivalenol, deoxynivalenol, and T2—produced by certain species of the fungus *Fusarium*, an organism (according to State Department experts) not usually found in Southeast Asia.

Eyewitness reports by Laotians claimed Soviet-built AN2 aircraft were used to spray and fire rockets carrying a toxic "yellow rain" on anti-Vietnamese village people. Since no facilities are known to exist in Southeast Asia capable of producing the mycotoxins found in samples from both Laos and Cambodia, and since (according to Department of State) the Soviet Union is known to pursue research on mycotoxins and maintain facilities for their large-scale production, circumstantial evidence strongly suggests that the Soviet Union is supplying mycotoxins for BW use in Southeast Asia (and possibly Afghanistan). TASS immediately labeled the charges "unfounded and false."[27]

Strong objections to the Department of State claims have been voiced. J. R. Bamburg, a toxicologist and recognized authority on T2, and M. Meselson, an acknowledged expert on chemical and biological warfare have questioned whether a single sample, collected without controls, stored and shipped under circumstances which could have resulted in its contamination, is indeed enough on which to base a meaningful accusation.[28] However, subsequent to Bamburg and Meselson's critical statements, the Department of State considerably strengthened its case by gathering under one cover all available evidence on the employment of chemical and biological warfare in the Indochina peninsula and Afghanistan.[29] The entire episode, as of this writing, seems to be a long way from being resolved and may cause further dissension between the two superpowers.

BW and an Interspecies Conflict

From the foregoing account it is clear that BW has not significantly affected the course of any human conflict. However, there has been a conflict which continues to this day, where the major determinant is BW. Fortunately, the conflict has not involved warring nations; rather, we find mankind opposing a tough and persevering adversary—the common rabbit.

Homo sapiens Versus Leporidae

Members of the rabbit family *Leporidae* have been hunted by man since the dawn of human history. Their meat has been a useful adjunct to man's diet and their fur useful for making articles of clothing. On the other hand, members of *Leporidae* can be extremely destructive to crops and pastures. The most destructive species is the common rabbit, *Oryctolagus cuniculus*. The part of the world where the rabbit has caused particular adverse ecological and economical damage is Australia.

Europeans began to settle the Australian continent at the end of the eighteenth century, but the rabbit is believed to have arrived in Australia by the clipper *Lighting* in 1859.[30] The skipper, Thomas Austin, liberated an unknown number on his estate in South Australia where they procreated at a prodigious rate. As they multiplied they spread, sometimes at the rate of 70 miles per year.[31] The rabbits soon became a serious pest, damaging crops and competing with livestock for grazing pastures. By the end of the century, despite all human attempts to limit their advance, rabbits had spread to the Indian Ocean, approximately 1100 miles from the point of their original liberation. As the enormity of the problem grew, management efforts were initiated on local and state levels to control their spread and numbers. The organizational effort to manage the rabbits need not concern us here; it is sufficient to say that despite all efforts, rabbits became entrenched throughout Australia's temperate zone, the area of fertile soils on which great farms abound.

Classical methods of management used in an intelligent, systematic fashion were sufficient to control the rabbit problem in individual cases. But general control at the commonwealth or state level was never accomplished, nor did it seem likely that such a goal was reachable. Eventually the commonwealth decided to put the entire question of rabbit control on a scientific basis. Help was sought from Australia's major scientific organization—the Commonwealth Scientific and Industrial Research Organization (CSIRO). In 1949, CSIRO began to discuss the possible use of biological control over rabbits and thus came to consider the myxoma virus as a suitable agent.[32]

The myxoma virus is a DNA virus of the poxvirus group. The myxoma is host specific—it will only attack four species of leporids, including the common rabbit. However, while the virus will cause localized fibromas (a tumor made up of fibrous tissue) in the other three species, the common rabbit will suffer from lesions spreading over the entire body, resulting in very high mortality rates. The virus is spread by a a variety of stinging insects, commonly by mosquitos but also by fleas, lice, mites, and ticks.

Research under the auspices of CSIRO soon uncovered earlier attempts to use myxoma virus, which had proven unsuccessful. Three attempts had been made in England, Denmark, and Sweden during the years 1936 to 1939. In all three cases the disease soon fizzled out. Previous small-scale attempts to introduce the disease in Australia during the year 1942, 1944, and 1950 had also failed, presumably because of the absence of an insect vector to disseminate the virus properly.

Despite previous failures, CSIRO scientists in 1950 carefully chose four separate sites to attempt the introduction of myxomatosis into large rabbit populations. These attempts also seemed headed for failure—at three sites the disease quickly appeared to have died away. Quite suddenly, however, farmers began to observe large numbers of sick rabbits around Balldale. By mid-February a huge area, covering south Queensland and north New South Wales, was affected.[33] Millions of rabbits became infected, resulting in a mortality rate in excess of 90 percent. Additionally, a large percentage of male rabbits who recovered suffered whole or partial infertility.[34] Epidemiological studies soon uncovered a relationship between disease transmission and the Murray-Darling river system. It was realized that unusually heavy rains had produced an extremely favorable situation for mosquito breeding resulting in a larger than usual mosquito population capable of carrying and spreading the myxoma virus. With the arrival of dry weather spells, the epizootic began to wane and by March 1951 most myxomatosis activity had ceased. Still, it was clear to scientific observers that the disease had become endemic. When spring came, there was a recurrence of the myxomatosis epizootic.

The results of using myxomatosis to control rabbits is impressive. The estimated 80 percent or greater reduction in numbers of rabbits since 1950 has meant farmers have been able to increase stocking rates of sheep and cattle because grazing pressure had eased on pastures.[35] The "rabbit droughts" (so called because otherwise fertile fields laid to waste by hordes of rabbits will resemble parched lands ravaged by droughts) do not occur any more. Hills bare for decades now are verdant. Forests previously in danger of destruction because rabbits were devouring all tree seedlings are now mending. Efforts by farmers to control damage caused by rabbits still continue, but no longer are as demanding on resources, and a substantial reduction in rural debt has resulted.[36]

One may at this point ask what the costs were of introducing myxomatosis to a land area where the disease had never previously occurred. From a retrospective view, the negative side effects have been minor. On the aesthetic side one can consider the unpleasantness of, at times, having thousands of decaying, disfigured rabbit carcasses littering the landscape, while thousands of other sick, unsightly rabbits mill about suffering obvious agonies. Only one major, perhaps inevitable, cost comes to mind—the development of resistance in rabbits to myxomatosis. Upon the initial introduction of myxoma virus the rabbit mortality rate ranged from 90 to 99 percent. By 1955, it became obvious the virus' disease potential had decreased; eventually the highest mortality rate that could be expected after inoculation was 75 percent.[37] The rate may continue to decrease, and the development of resistance is of serious concern in a country as dependent on agriculture as Australia.

Other costs are insignificant, for example, the direct financial costs of producing and dispensing virus inoculum is small when compared to the savings realized by farmers. Therefore, the cost-benefit ratio in this case of employing biological control measures against a pest has been disproportionately favorable.

Biological Control as a BW Device

The use of myxomatosis as an instrument of destruction has been described in detail because the episode presents a model for a future human-versus-human conflict where one side decides to employ an efficacious, controllable BW agent.

From the use of myxomatosis and the history of BW presented earlier, some points can be made and lessons drawn:

1. When large-scale national efforts have been made in the past to develop and produce BW weapons, detection has proven difficult, if not impossible. For example, only toward the end of World War II did Allied intelligence services detect Japanese BW activity, and they were unable to ascertain its scope. It may be concluded that a sizable national effort may be undertaken for research and development of BW systems without other nations becoming aware of the effort.
2. Past employment of BW in human conflict has been on a very small scale and has most probably made insignificant contributions to battle.
3. It is very difficult for a victim to "prove" BW aggression. The Chinese admitted as much:

> On account of its very nature, the use of biological weapons is an act exceptionally difficult to prove. Perfect proof might require, for example, that an airplane be forced down with its biological cargo intact and its crew prepared to admit their proceedings forthwith. Obviously, this would be a very unlikely occurrence for many reasons.[38]

Of the historic examples presented, only in the case of Japan is there acceptable evidence of sufficient quantity and quality to ascertain that the Japanese developed and used BW prior to, and during, World War II.[39]

This point could be taken one step further. A nation (or community) may become a victim of a BW attack and never know it. Such an event could take place if an epidemic caused by a BW agent is thought to have a natural basis. If an attack is suspected, proving it to the international community could be so difficult as to preclude the possibility of demonstrating a legally acceptable cause-effect relationship.

4. It is very difficult for a party accused of having waged BW to establish innocence, although strong presumption of innocence is possible if an accused nation can demonstrate the absence of BW R&D facilities and is fully committed to open research.

5. The use of BW by man against rabbits may demonstrate the possibility of employing similar means effectively in human conflicts, provided certain conditions are present. First, the target population must be clearly identifiable through its biological characteristics, and one characteristic should be location at the end of an ecological food chain. If this is the case, the elimination of a target species or genera would have a small impact on the biosphere. Next, the BW agent chosen must possess an extremely narrow host range, and preferably be host-specific, while the attacked population should be unable to protect itself easily through the use of such means as vaccines. Finally, the effects of using a BW agent must be largely predictable, especially the negative side effects, and the action of the agent must be controllable by its employer.

In Australia, the rabbit was an easily identifiable target positioned at the end of a food chain (predation by man and other animals was insignificant). The agent employed was host-specific, its effects were well known and predictable, as was the lack of significant side effects. Though an effective vaccine against myxomatosis exists, the rabbits were, of course, in no position to use it. The action of the agent was somewhat controllable by taking advantage of weather conditions, seasonal changes, and land characteristics. The use of myxoma virus remains for these reasons an example of the highly efficient use of BW by man and an ominous portent of the harm the use of an appropriate BW agent could cause to an unsuspecting or exposed target population.

Characteristics of BW Agents

In a report prepared in 1968 for the Conference of the Eighteen-Nation Committee on Disarmament, the UN Secretary General stated four important characteristics of a BW agent: It should be able to be produced in quantity, be capable of ready dissemination in the face of adverse environmental factors, be effective regardless of medical countermeasures, and be able to cause a large number of casualties.[40] In Table 1, I suggest a list of attributes modified in light of subsequent knowledge.

Table 1. Attributes for an Efficient BW Weapons System (in descending order of importance).

1. High virulence coupled with high host specificity
2. High degree of controllability
3. High degree of resistance to adverse environmental factors
4. Lack of timely countermeasures
5. Relative ease of camouflaging BW agent

High virulence[41] *coupled with high host specificity* means that the agent should have the capability of severely debilitating or killing the targeted host without having a spillover effect on other species. A *high degree of controllability* means that the employer of a BW agent should be able to target the agent with high accuracy, and when the preplanned objective has been reached (or nearly reached), the agent should die, disappear, or become inactivated, again on a preplanned basis. *High resistance to adverse environmental factors* means that the BW agent, while in transit, should be able to survive temporary adverse factors that, under usual circumstances, would inactivate a related "wild" type of disease agent. Adverse environmental conditions could include dryness, high altitude, sunlight, cold, heat, and so on. *Lack of timely countermeasures* means that the BW aggressor should employ a weapons system capable of causing extreme damage before efficient defenses are activated. Against every BW weapons system so far known there have been effective countermeasures available, or such means can quickly be developed.

Camouflaging the agent means that the attacked party should be unaware of being the target of aggression. Proper planning could allow an aggressor to mount an attack with an organism normally present in the victim's environment. For example, there exist in the American Southwest and Mongolia large pools of animals harboring plague. Potentially, an epidemic of plague could start in these areas at any time. The introduction of a particularly virulent form of plague into, or proximal to, these areas by an aggressor would be soon apparent, but it would be doubtful that the ultimate source could be detected. If suspicions of BW aggression arose, the accused party would still be in a position of explaining away the event as a natural phenomenon. (On the other hand, a BW aggressor may not care if the fact becomes known. If so, attribute 5 becomes unimportant.)

There are a number of agents that have been examined for potential BW use against humans, but none possesses all five of the attributes discussed above. Tables 2 through 5 show attributes and ratings for myxomatosis and for three BW agents potentially useful in human conflicts. Figure 1

Table 2. Myxomatosis.

Attribute	Rating
1. The agent is highly virulent and will attack only the rabbit.	Highly acceptable
2. The agent may be partially controlled by seasonal use and by a thorough consideration of geographical and environmental factors. However, due to its species specificity, this attribute is not so important when considering the myxoma virus.	Acceptable
3. For the purpose of decimating rabbit populations, the agent is stable and easily dispensed.	Acceptable
4. No countermeasures are available to the rabbits except the development of resistance to the agent over a period of time.	Highly acceptable
5. In this instance, attribute 5 of Table 1 is not applicable; however, the first Australian epidemic could conceivably have begun naturally.	Acceptable

Table 3. Anthrax.

Attribute	Rating
1. The agent is virulent to humans. Though not specific to humans, its host range is sufficiently narrow so that spillover is limited.	Acceptable
2. The agent may strike any human; therefore, a backlash may affect the aggressor (unless immunized). Further, the agent may remain in the environment for a long period of time in the form of spores in dirt and dust.	Unacceptable
3. The agent may easily be rendered resistant to environmental influences.	Acceptable
4. Though countermeasures exist against the agent, its virulence (high case fatality rate) may preclude the timely application of those countermeasures.	Acceptable
5. The origin of an anthrax (or plague) epidemic could, under certain circumstances, be camouflaged. Nevertheless, an epidemic in a developed nation would be highly unlikely.	Conditionally unacceptable

gives a qualitative comparison of the "ideal situation" of myxomatosis with the same three agents.

The number of disease agents surveyed here is small, but I believe that expanding the survey would only strengthen the conclusion: The major deterrent to BW has been, and remains, the lack of controls over the agent once it has been released. So far, the possibility has always existed that the soldiers and population of the BW aggressor could themselves become victimized by the released agent. As a result of that uncertainty, nations have been loath to use BW.

Table 4. Viral Encephalitis (Eastern Encephalitis).

Attribute	Rating
1. The agent is highly virulent and specific.	Highly acceptable
2. The agent attacks humans indiscriminately and is carried by the mosquito or other biting insects. Thus, nontarget populations may also be attacked.	Unacceptable
3. The virus particle can be adequately protected by enclosing it within an insect.	Acceptable
4. An effective vaccine against the virus exists but the organism's virulence may preclude its effective use. Therefore, unless the defender anticipates the use of viral encephalitis in a conflict, a vaccine is useless. Furthermore, no cure exists.	Acceptable
5. The potential for outbreaks of this disease exist in many parts of the world.	Acceptable

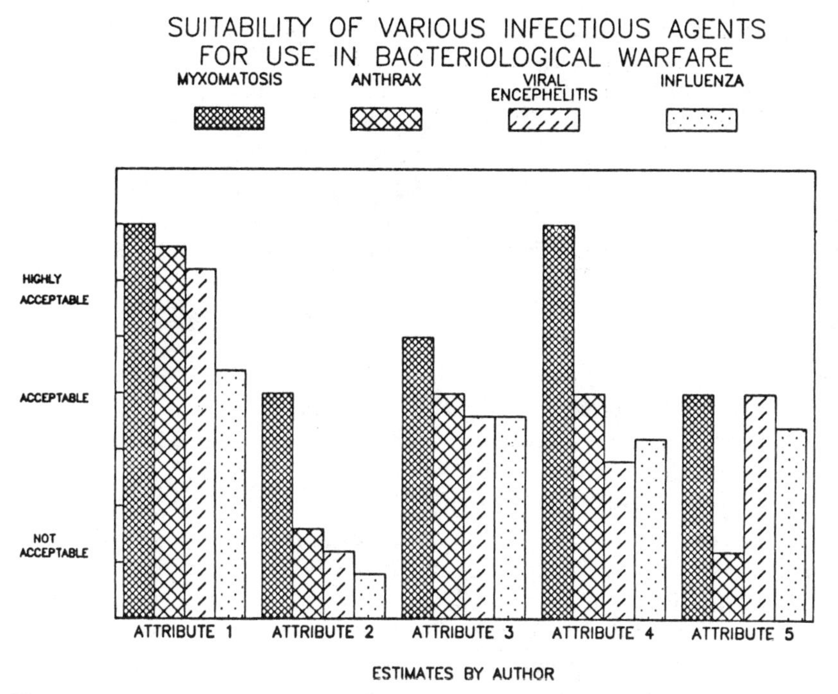

Figure 1. Suitability of various infectious agents for use in bacteriological warfare.

Table 5. Influenza.

Attribute	Rating
1. The agent usually does not cause a virulent, life-threatening disease, but it is rather specific to humans. The agent may be used in a strategic sense, that is, to cause economic damage by prostrating large numbers of a population.	Acceptable
2. The virus may spread across the world with lightning speed, indiscriminately striking people everywhere.	Unacceptable
3. The virus, when part of an aerosol, is resistant to environmental stresses.	Acceptable
4. The potential exists for producing an effective vaccine within a few months after the virus has been first detected. Nevertheless, before a vaccine comes into use, a large amount of damage may occur.	Acceptable
5. The disease is easily camouflaged.	Acceptable

Recombinant DNA Research and BW

With this background, I can now discuss manipulations of bacterial and viral genomes that rDNA techniques make possible and that have implications for the development of BW agents. At the present time although rDNA techniques allow scientists to effect change in only a small number of genes, it is possible to accomplish three types of modifications: (1) to increase the resistance of bacteria to antibiotics, (2) to increase or enhance various virulence factors, including the ability of bacteria to produce toxins, and (3) to rearrange viral genes that code for surface antigens. The reader is reminded that experiments to accomplish any of the above modifications were proscribed by the Asilomar conference and the original NIH guidelines.

Resistance

Genes that code for resistance to antibiotics are carried by the R factor plasmid. Bacteria are able to pass R factors to one another by the process of conjugation, sometimes crossing interspecies barriers. The pressures brought to bear on bacteria by the use of antibiotics have led to the development of large antibiotic-resistant populations. As the incidence of resistance grows, the costs involved in treating infections plaguing mankind increase explosively. Costs may be expressed in financial terms, in increased suffering by sick persons, and in increased morbidity and mortality rates.[42]

The transfer of R factors from one organism to another can be accomplished either through "classical" means or through rDNA techniques.[43] The latter enable molecular biologists to transfer R factors relatively easily between *unrelated* bacterial species—a unique achievement and one that multiples the possibilities of creating antibiotic-resistant organisms. Given such organisms, the following scenario may be presented.

Nation *A* is at war with nation *B*. *A* knows about the usual medical practices in *B*, including the accepted means employed to treat dysentery. In *B* the first choice antibiotic for treating dysentery is ampicillin, the second is streptomycin, and the third is chloramphenicol. *A* also knows that the facilities for producing another group of antibiotics—cephalosporins—is poor or nonexistent in *B*. *A* proceeds to produce large quantities of *Shigella* (the dysentery bacillus) resistant to the first three antibiotics but sensitive to the fourth. The resistant *Shigella* is thereupon introduced into *B*'s water and food sources to the fullest extent possible, while at the same time *A* makes certain its own military and civilian medical services have a full supply of cephalosporins on hand.

A similar, purely speculative scenario could be based on the incomplete facts from the Sverdlovsk anthrax outbreak: possibly an accident involving a BW system had taken place at Sverdlovsk. The allegedly high casualty rate experienced by Sverdlovsk's population could have been due to the disease having been caused by an extremely resistant form of *B. anthracis* developed specifically for BW. Thus, the primary drug of choice, penicillin, and other antibiotics would have been useless, and the epidemic was not controllable until BW specialists communicated their information to the appropriate public health agencies, who were then able to supply efficacious antibiotics to the stricken populace.

Virulence and Toxins

Studies of enteropathogenic *E. coli* (which causes inflammation in the intestinal tract) have shown these forms to routinely carry five or more distinct plasmid species not present in nonpathogenic *E. coli*.[44] Individual plasmids may code for antibiotic resistance, for the production of enterotoxins, and for colonization or adherence antigens. Some plasmids may code for virulence factors capable of affecting several animal species, while yet others code for species-specific factors.[45] Similar plasmids that code for virulence factors exist in other bacterial species. Examples of virulence factors coded for in plasmids, or similar extrachromosomal elements, include enterotoxin B found in strains of *S. aureus* responsible for food poisoning, the exfoliate toxin produced by *S. aureus* causing toxic epidermal necrosis in newborns, and an as yet unnamed factor enhancing tissue invasiveness by *Yersinia enterocolitica*.[46]

As with R factors, virulence factors can be transferred between related species using both classical and new techniques, but the transfer between unrelated species can only be done using rDNA techniques. For example, it is now theoretically possible to transfer the gene coding for botulin toxin production from *Cl. botulinum* to *E. coli*.

For the purpose of BW, steps to enhance virulence may be taken for at least four reasons:

1. To increase mortality rates. In nature a parasite whose action is so damaging as to cause the host's demise is generally an unsuccessful life form, since the host's death will inevitably result in the parasite's death. For BW purposes this ending is quite favorable, because the quick elimination of both the attacked party and the destructive agent adds a measure of control over the process by the attacker. Therefore, enhancing virulence to the point where the altered organism is extremely deadly could supply a heretofore missing element of control.

2. To increase morbidity rates. An attacker may find it useful to debilitate large numbers of his adversaries. One means would be for an attacker to insert several types of plasmids into a BW agent, for example, those coding for both virulence and antibiotic resistence. The attacked party would face an extremely difficult situation by having to respond both quickly and with the correct drugs (which may not be readily available).

3. Allow the attacker a greater freedom to choose among targets. It has been shown that certain plasmids may have species-specific effects. An attacker may be in a position to take advantage of plasmid specificity. For example, a nation (or area within a nation) dependent on raising pigs could

possibly be economically devastated by the dissemination of an appropriate plasmid-containing enteropathogenic *E. coli* striking only pigs.

Before leaving this section, another facet of toxin production should be mentioned. Some of the most toxic substances in existence are biological toxins (biotoxins), some of which may be over 1000 times as powerful as cyanide (Figure 2). The exceedingly high degree of toxicity of a biotoxin, as well as a new use for these agents, may have been demonstrated by a series of events in 1978. It is thought that a biotoxin may have been used in attempts to assassinate two Bulgarian defectors, one of whom died. In both instances, pellets were introduced under the skin. In the first incident it is probable that an assailant used an airgun to fire a pellet into Vladimir Kostov's back while he was in a Paris Metro station. The pellet, making only a shallow penetration, was quickly removed with minor pain and no lasting consequences. The other Bulgarian, however, was not as fortunate. Three weeks after the first incident, Georgi Markov was prodded by an umbrella while walking in London. Four days later he died of cardiac arrest. At the autopsy a pellet, which proved to be identical to the one recovered from Kostov, was removed. Upon examination, the pellets were measured to be 0.17 cm in diameter and each contained two cavities capable of containing 0.4 mg of solid material.[47] Unfortunately, one can only speculate on the most probable cause of death. If one consults Figure 2, one finds that the only toxic substances sufficiently lethal to kill a 70 kg person when administered in a 0.4 mg dose are biotoxins and gamma emitters. No trace of radioactivity was found in the pellets. On the other hand, once a biotoxin has dissolved and been distributed throughout the body, no method exists at present for its detection. Therefore, a possibility exists that a biotoxin was used in the two attacks.[48] In any case, biotoxins, being both lethal in extremely small doses and difficult to detect, offer terrorists and assassins alluring possibilities.

In particular, the transfer of toxin-coding genes from, for example, *Cl. botulinum* to *E. coli* is possible with today's technology (though such a project remains forbidden since Asilomar). Engineering *E. coli* to produce botulinum toxin could eliminate the present major problem of production, namely, the requirement for an absolute anaerobic environment. Botulin producing *E. coli* (*E. coli* Bot+) could be liberally added to water or food supplies in order to reach the widest scope of targets. Most probably, the altered *E. coli* would not be able to survive competition from wild *E. coli*, thereby ensuring the elimination of *E. coli* Bot+ after a few generations—and also after it had been given a chance to eliminate large numbers of people. Of course, the possibilities of designing and manufacturing various toxin-producing bacteria are not exhausted by this one example; many other combinations could be equally attractive to the BW aggressor. After all, a wide variety of proteins can theoretically be produced by bacteria or yeast through the employment of rDNA techniques, the most important factor

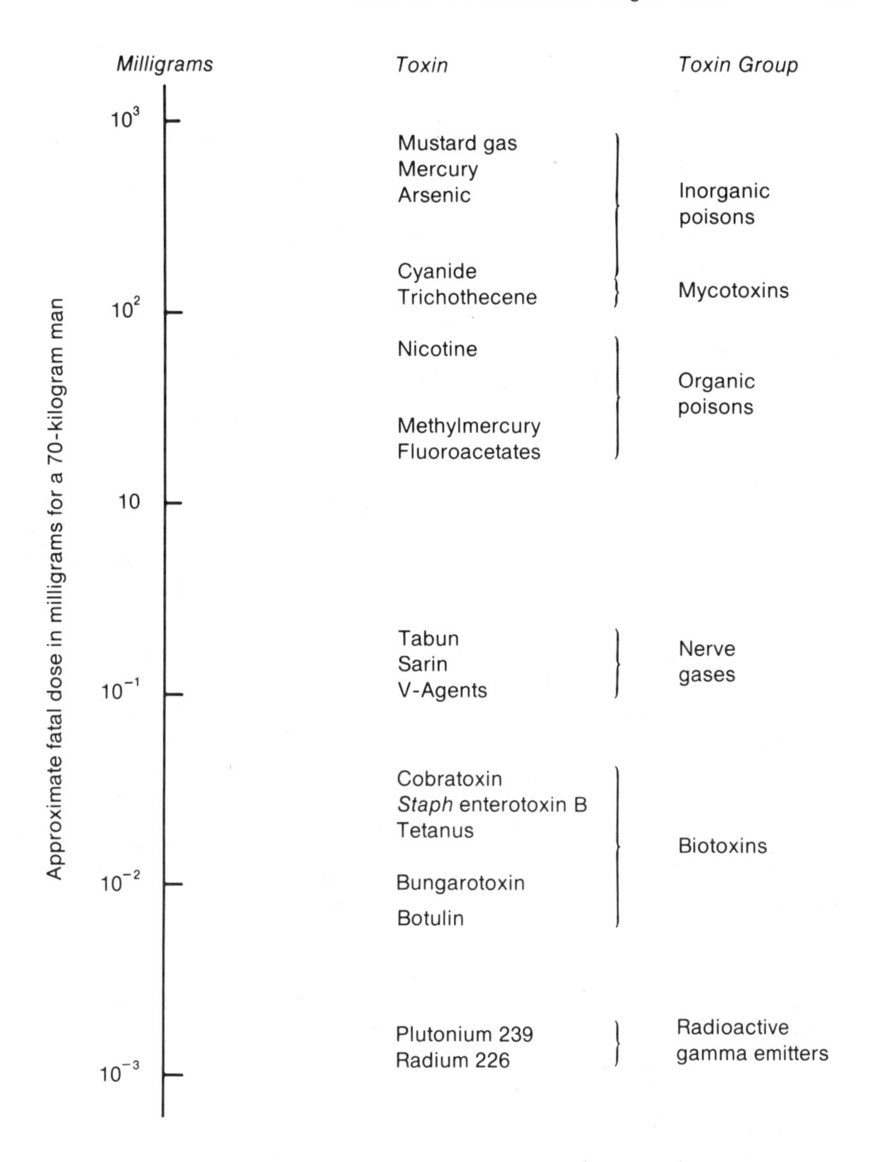

Figure 2. A comparison of toxic doses of various poisons. Adapted from J. Emsley and D. Pallister, "Bulgarian Brolly Baffles Germ Warfare Boffins," *New Scientist* 80 (1978): 92; and United Nations, Review Conference of the Parties to the Convention on the Prohibition of the Development, Production, and Stockpiling of Bacteriological (Biological) and Toxic Weapons and on Their Destruction, *Final Document*, UN Document BWC/Conf. 1/10, 21 March 1980, p. 13.

being whether the production is controlled by one gene or many. A process coded for by a very few genes is now accomplishable—processes controlled by many genes may be manipulated tomorrow (or the day after).

Viral Genes

So far I have discussed only the use of rDNA techniques in manipulating bacterial genes located in plasmids, the procedures easiest to perform. However, viral genomes have been taken apart by researchers and the individual genes cloned in bacteria. Today's techniques allow for rearranging viral genes that code for surface antigens; thus it is possible to alter the antigenic makeup of a virus so that antibodies now present in populations would be ineffective against it. A future BW researcher could conceivably alter the antigenic makeup of a flu virus to resemble the deadly form that in 1917 killed many millions of people throughout the world.[49] Before releasing the "new" flu, sufficient quantities of effective vaccines could be manufactured and dispensed to protect favored populations.

The foregoing discussion relating the possibilities offered by rDNA research for the design, production, and dissemination of BW agents for warfare or terrorism has been limited. It is easy to imagine untold numbers of other combinations using various types of microorganisms, biotoxins, and pathogenic properties, all holding many promises for the future, enterprising BW armorer.

If we now couple the possibilities rDNA techniques may offer to BW with the promises rDNA research holds for peaceful uses in medicine, agriculture, industry, and energy (see Appendix), we may make a few conjectures. First, recent advances have simplified many rDNA techniques to a level that allows graduate students in molecular biology, biochemistry, or other scientific disciplines to perform rDNA experiments with confidence. It is not unreasonable to expect that in a short time these same experiments will be performed by undergraduate-level laboratory workers.[50] In effect, this progression of events means that more persons working under less demanding circumstances could perform work useful for BW. Second, I contend that advances in the field of microbiology together with the potential of rDNA research will soon make it possible for scientists to decrease the technically objectional aspects of BW (principally the lack of controllability) so that its use could be seriously considered by national leaders or terrorist groups. Third, the time frame for these developments is, in my opinion, fairly short, perhaps five to ten years.

If these conjectures come true, the following developments may come to pass:

1. Large-scale production of "new" organisms useful in BW could be undertaken successfully.
2. Organisms will be designed to remain viable and virulent in transit or when stored to a greater extent than previously possible.
3. Organisms will be designed for easy dissemination in a manner calculated to cause the most damage.
4. Organisms will be designed to resist, with minimum loss of virulence, adverse environmental factors after release.

5. Antigenically unique organisms will be designed and manufactured so that populaces have little or no natural immunity against them. Their unique character will preclude timely prophylactic activity, such as vaccinating populations against the BW agent.
6. If the BW agent is of bacterial origin, it may be designed to resist most, or all, known antibiotics.
7. Specific immunotoxins (a monoclonal antibody linked to a potent toxin molecule) will be developed to attack certain subpopulations selectively.
8. The attacker will be able to retain control over the BW agent by designing it to cause a maximum effect on the host in a short time, then to die off after a previously determined number of cell divisions (controlled senescence) without causing any secondary infections; or to be bound by a narrow set of environmental factors, for example, if the temperature falls under 15° Celsius the organism would be unable to survive.
9. In the long term it may be possible to design an organism to be species specific to *Homo sapiens* (or specific to a human subset—for example, to only those with type O blood) by building into it one or more of the control mechanisms suggested in item 8 above. Its employment would cause few, if any, short- or long-term detrimental biospheric consequences.

Anti-BW International Legal Instruments

Given the possibilities that rDNA research may offer to BW, are there legal constraints in existence to prevent BW? In a previous paper, I have described a series of historic measures that eventually culminated in the 1972 Biological Weapons Convention (BWC).[51] Only one of these—the 1925 Geneva Protocol—need be discussed here. I will then consider the BWC, and will follow with some comments on international terrorism.

Geneva Protocol

The Protocol for the Prohibition of the Use in War of Asphyxiating, Poisonous, or Other Gases, and of Bacteriological Methods of Warfare was signed at Geneva on 17 June 1925.[52] The aim of the protocol was to extend the already existing prohibition against the use of toxic chemical agents in war to include bacteriological weapons: "That the High Contracting parties . . . agree to extend this prohibition to the use of bacteriological methods of warfare and agree to be bound as between themselves according to the terms of this declaration."

Though the Protocol was widely accepted, the United States did not ratify the treaty until 10 April 1975, subject to reservations. More than forty other nations attached reservations to their acceptances, usually reserving the right to retaliate in kind if they became targets of a chemical or bacteriological attack.[53] Among those reserving the right of retaliation are Great Britain, China, France, USSR, and the United States. With time another severe restriction became apparent—nations are free to develop offensive biological weapons though they are proscribed from using them.

The BW Convention

During a 1968 meeting in Geneva, the British representative to the Conference of the Eighteen-Nation Committee on Disarmament introduced a draft BWC calling for an attempt to redress the shortcomings mentioned above and for separating the chemical from the biological area for the sake of negotiation. These ideas were generally acceptable and so were included in the disarmament committee's report to the United Nations General Assembly (UNGA). In turn, the UNGA passed a resolution that called on the Secretary General to prepare a report on chemical and biological warfare weapons and the possible effects of their use.[54] Less than a year later, the report, prepared by a group of experts, was distributed to all UN member states. Perhaps a modicum of urgency was given to the recipients by the consultant experts' aspirations:

> It is the hope of the authors that this report will contribute to public awareness of the profoundly dangerous results if these weapons were ever used, and that an aroused public will demand and receive assurances that governments are working for the earliest effective elimination of chemical and bacteriological (biological) weapons.[55]

The British initiative soon led to a Convention signed by the U.S. and Soviet representatives on 4 April 1972, followed by representatives of more than eighty other countries. The BWC, done in Washington, London, and Moscow on 10 April 1972, was not entered into force for the United States until 26 March 1975.[56] The convention is the first, and so far the only, international instrument that calls for the elimination of an entire weapons system:

> Each party to the Convention undertakes never in any circumstances to develop, produce, stockpile or otherwise acquire or retain: (i) Microbial or other biological agents, or toxins whatever their origin or method of production, of types and quantities that have no justification for prophylactic, protective, or other peaceful purposes; (ii) Weapons, equipment or means of delivery designed to use such agents or toxins for hostile purposes or in armed conflict. (Article 1)

Each nation, upon signing the Convention, promised to destroy all existing BW stocks within nine months (Article 2) and adopt national measures to prohibit public or private concerns from performing any facet of BW development (Article 4). Any nation suspecting another to be in violation of Convention provisions can lodge a complaint with the Security Council. In keeping with the provisions of the UN Charter, the Security Council may undertake an investigation of the accusation, and all involved states are obliged to cooperate with the investigation (Article 6). Though the BWC is limited to biological and toxin weapons, a link between biological and chemical warfare (CW) is found in Article 9, which affirms the need to prohibit CW and asks nations to continue negotiating for that purpose. Article 10 contains an interesting provision for the continuous exchange of ideas and information between nations on the use of biological processes for peaceful purposes. The exchange is to be accomplished by two methods. First, all signatory states have the right to participate in "the fullest possible exchange of equipment, materials, and scientific and technological information for the use of bacteriological (biological) agents and toxins for peaceful purposes." Second, those states in a position to help others shall do so in a manner whereby biological processes would be useful in the prevention of disease "or for other peaceful purposes." Recognizing the volatility of the field, Article 12 calls for periodic reviews of the Convention in order to "take into account any [relevant] new scientific and technological developments." The reviews are to take place at five year intervals after the Convention entered into force; the first took place in March 1980.[57]

Shortcomings of Anti-BW Legal Instruments

As with any multilateral treaty, the BWC has several shortcomings and omissions. Beginning with the most serious, major inadequacies of the BWC are these:

Verification. Aspin has described the importance of vertification to arms control agreements:

> The keystone of an international arms-control agreement is the ability of each side to make sure the other side abides by it. Without adequate verification of compliance, agreements such as the bilateral strategic-arms pact between the US and the USSR are bound to collapse.[58]

Verification of the understanding reached in SALT II, for example, can be done without actual on-site inspection by using what Aspin calls "national technical means." These include direct surveillance of each other's territories using space satellites, radar, and high-altitude aircraft; and indirect means using seismological devices, air sampling, electronic tracking of aircraft, missiles, and submarines, and so forth. According to Aspin, these means of verification have proven adequate to both the United States and the USSR, therefore, it is most probable that SALT II provisions will be followed through 1985 as agreed upon.[59]

There is no reason to believe verification should be less important to BW and CW arms control agreements than it is to SALT II; however, no means of verification are included in either the Geneva Protocol or the

BWC. Further, no adequate "national technical means" exist to substitute for the lack of formal verification instruments in monitoring national BW or CW activity. For example, each nation, upon adopting the BWC, promises to destroy all its BW stores within nine months. Then, the nation certifies its compliance, but no instrument exists for verifying the alleged compliance. Similarly, in contravention to Article 1, a nation may begin a research and development effort to produce BW systems in secret. As the Japanese BW program in World War II demonstrated, the probability of detecting such illicit activity is low in the absence of verification procedures.

Processing Complaints. If one nation suspects another of violating any of the BWC provisions, the first nation may lodge a protest with the Security Council (Article 6), accompanied by all possible evidence. The Security Council may then initiate an investigation "in accordance with the provisions of the Charter of the United Nations." This means that any permanent member can use its veto power to stymie the proposed investigation. No other means exist within the BWC to carry out an investigation of alleged wrongdoing.

National Legislation. When a nation signs an international treaty, it is obliged to bring its laws into compliance with the terms of the treaty. The BWC spells out that obligation (Article 4); as of 1982 very few nations had complied. One nation that did was the United Kingdom—the Biological Weapons Act of 1974 extended the requirements of the BWC to all English citizens.[60] The United States, on the other hand, has not complied.[61] Lack of compliance may mean that an industrial concern is legally able to produce substances useful to BW within the noncompliant nation's border.

Peaceful versus Warfare Use. The peaceful development of biological agents is permitted under Article 10(2). However, it is entirely possible to develop a product for peaceful use and later apply that same product in a harmful manner. For example, in 1979 the U.S. Army Medical Research Institute of Infectious Disease asked NIH's Recombinant Advisory Committee (RAC) for permission to engineer *E. coli* to produce the *Pseudomonas* exotoxin A.[62] This toxin has been studied under the Army's CBW program as a possible BW agent. However, the Army is also interested in the toxin because *Pseudomonas* is important as a complicating factor in burn wounds.[63] Permission for the Army to perform the experiment under P_1-EK_1 conditions was granted by RAC in March 1980.

Another example may be more pertinent. A component of the "yellow rain" discussed earlier is the fungal toxin diacetoxyscirpenol (commonly called anguidine). The substance, though extremely toxic, has shown antitumor activity in mice and hope was held that it might prove helpful against human rectal and colon cancers.[64] However, the yellow rain episode indicates anguidine could also be considered a potential BW agent. If production of the agent began under proprietary or other restricted circumstances, an outside observer may have difficulty deciding whether the production of anguidine was in violation of the BWC. Such a decision could be further complicated by inability to verify end use of the toxin and by the lack of conforming national legislation as called for by Article 4 of the BWC.[65]

Other developments in biotechnological techniques have brought added complications not considered by either the founders or reviewers of the BWC. The Defense Advanced Research Projects Agency (DARPA), part of the Department of Defense, is responsible for assessing future trends in research having possible applications for defense. A subunit within DARPA, the Advanced Bio-Chemical Technology Program, has reportedly asked for bids from researchers to advance futuristic concepts of applying biotechnological means to detect BW and CW agents.[66] The use of monoclonal antibodies may permit the construction of ultra sensitive, highly specific detection devices useful for, among other things, ferreting out BWC agents; discovering underwater weapons platforms; detecting pollutants; monitoring the health status of targeted persons; and tracking dope smugglers and manufacturers. These activities, being of a defense nature or undertaken for peaceful purposes, are certainly permitted under the BWC.

Another problem could arise with the use of proteins as semiconductors. These so-called biochips could, in the long term, replace the present semiconductors, made largely of silicon, for certain applications in computers and microprocessors. A possible problem arises if biochips integrated into circuits of computers and microprocessors are part of submarines, mines, missiles, aircraft, and other instruments of war having offensive attributes. Article 1 of the BWC proscribes the use of biological agents"for hostile purposes or in armed conflict." Does the use of biochips under such conditions contravene international law? The area is murky and the problems are as yet ill-defined.

New Processes. As the field of biotechnology expands, it offers attractive possibilities to the potential BW manufacturer, especially as a result of scale-up. Most likely, continuous fermentation procedures will be used to achieve massive propagation of bacteria and yeasts necessary for production of commercial products. Similar procedures could be employed to manufacture agents and products useful in BW. So, if such a path were chosen, today's techniques would enable enormous quantities of disease agents, biotoxins, or other biologically active substances to be produced in short order.

The cost of building, equipping, and staffing a rDNA laboratory to develop and produce BW systems in not exactly known, but an estimation has been made of the cost of establishing an international center for genetic engineering and biotechnology (Table 6).[67]

If these estimated costs are accurate, it seems possible for a nation to develop a respectable rDNA research and development capability and run it for five years, for not much more than $50 million (assuming the scientists able to run the facility are available). The estimated costs of developing nuclear weapons is in the range of billions of dollars, so, after comparing costs, a national decision maker may find the BW route to be the most attractive.[68] With a concentrated effort, utilizing both classical and rDNA

Table 6. Cost Estimates for Laboratory Development and Production of rDNA Products (costs at 1981 prices in U.S. dollars, ref. 67).

A. Fixed costs (excluding buildings, land, and utilities; including laboratory equipment and materials):	
1. Molecular biology and biochemistry departments	$ 2,600,000
2. Microbiology and molecular genetics departments	600,000
3. Advanced biotechnology department including pilot plant	4,650,000
4. Bioinformatics department	360,000
5. General services	720,000
B. Operational costs for 1 year for a staff consisting of 1 director, 1 deputy director, 30 scientists, 30 technicians, 20 clerical workers, and 15 manual workers	4,822,000
C. Other expenses	2,000,000
Total	$15,752,000

techniques, the potential power thus available could challenge the nuclear arsenals possessed by the superpowers.

Life Forms Made from Inorganic Chemicals. The possibility that biological products and, possibly, life forms manufacturerd entirely from inorganic starting materials may not come under the provisions of the BWC has been considered by Callaham and Tsipis.[69] The question is still unsettled, though I believe, as do Callaham and Tsipis, that an agent exhibiting vital phenomena characteristic of life would unquestionably fall under the provisions of the BWC. Problems that could arise out of the new processes and agents were discussed during the BWC Review Conference held in Geneva in March 1980 (see below).

If the BWC is so weak, why have nations so far eschewed BW? The following considerations, combining practical difficulties with societal strictures, may in the past have prevented the deployment of BW:

1. Lack of controllability of available BW agents.
2. Lack of stability, which causes problems with delivery and storage.
3. Lack of virulence upon dissemination due to the incapacitating action of the physical and biological environment.
4. The existence of countermeasures, either prophylactic (vaccines) or therapeutic (antibiotics), since most of the disease agents considered for possible use as BW agents in the past have been well known to health authorities the world over. No one nation has ever had a dominant position vis-à-vis other nations by achieving a unique BW capability; therefore, there has never existed a clear-cut advantage for any nation to employ BW.
5. The value of nuclear weapons as a means of mass destruction, which has so far outweighed BW weapons effects; therefore, the nuclear powers have had little incentive to develop BW.

6. The popular response to BW in democratic societies, which has been repugnance. Leaders of democratic nations find it difficult at any time to go against public opinion, therefore, BW employment has so far been proscribed. Public opinion in totalitarian states may count for less but still remains a factor.
7. The unpredictability of BW agents and the possibility of complicated negative consequences occurring after BW use. This may have led to the establishment of a *pactum turpae** between nations preventing BW employment.[70]
8. International law measures, which seek to prevent employment of BW. Specific measures already discussed include the Geneva Protocol and the BWC (the Nuremburg Principles are also relevant here.[71])

So far these considerations have prevented BW deployment and use. The possibility of most of the practical difficulties becoming modified through the use of rDNA techniques are discussed above. Other influences, for example, the existence of extreme hate or fear, could negate societal strictures, thus upsetting the *pactum turpae* and negating international law. If inhibitions were removed, rDNA research could offer the technical means for overcoming limitations so far present in BW systems.

The Review Conference on the BWC

In March 1980, the first review of the BWC was held. During the first review, discussion was limited strictly to scientific and technological developments pertinent to BW. The five developments discussed included rDNA techniques, "new" infectious diseases, toxins, recent industrial procedures, and microbial pesticides.[72] The four major conclusions reached by the review committee were these:

1. Acceptance by nations of the BWC has not hindered peaceful research and development activities.
2. Although new scientific and technological development could permit the development of microbial and other biological agents or toxins with enhanced military utility, these new agents are unlikely to improve upon known agents to the extent of providing compelling advantages for illegal production or military use in the foreseeable future.[73]
3. The language in the BWC is broad enough to cover any new biologicals that could be manufactured with rDNA techniques.
4. The development of new and improved processes in the biotechnological industry as well as general scale-up add to nations' capabili-

*Pactum—an agreement; turpis—base, mean, vile, infamous, unlawful.

ties to produce BW substances. This situation is not unique to the biotechnological field since other civilian capabilities, for example, the chemical industry, can be diverted from peaceful use. In any case, the new and expanded capabilities "do not appear to alter substantially capabilities or incentives for the development or production of biological or toxin weapons."[74]

Evaluation

On the basis of the information presented earlier and the findings of the UN's BWC Review Conference, one can see four possibilities:

1. That new techniques, including those from rDNA research, have not changed the BW field.
2. That new techniques have made it possible to enhance BW systems but the changes are quantitative, not qualitative. These changes do not demand a political and legal rethinking of presently existing control measures.
3. That techniques introduced by rDNA research have changed BW in a qualitative manner by making possible the design and manufacture of entirely new BW systems. Decision-makers, therefore, must reconsider the entire field of BW and, as necessary, develop approaches to meet the new challenges.
4. That it is too early to assess, with any degree of certainty, the potential that rDNA research offers to BW.

In contemplating the four alternatives, the reader must surely agree that alternative 1 can be a priori eliminated—scientists do now have the previously undreamt of ability to genetically alter and control microorganisms.

Alternative 2, as mentioned earlier in this chapter, is the one favored by the BWC Review Conference. A strong argument can be made to support its conclusion, since possibilities of greatly enhancing several characteristics in pathogens deemed desirable in BW systems—for example, virulence, resistance to both antibiotics and stressful environmental factors, controllability, and so forth—do not present any new scientific challenges to the health and medical professions. Thus present international and municipal legal and political measures are more or less adequate.

The alternatives 3 and 4 overlap considerably. It is unlikely that entirely new microorganisms have so far been created using rDNA techniques. Nevertheless, in this chapter, possibilities of designing and producing new strains of microorganisms specifically for use in BW have been suggested. For example, it is conceivable with today's knowledge to design bacteria to be resistant to all known antibiotics; to endow nonpathogens with the ability to produce immensely powerful toxins; or to engineer viruses and bacteria to die off quickly after having caused great, immediate damage—thereby giving the BW user a degree of control over BW never before attained.

In my view, the prudent approach by the international community is to accept alternative 4 and, by implication, to consider alternative 3 seriously. Evidence presented throughout this volume indicates the high probability of dramatic changes taking place throughout the entire peaceful field of biotechnology. Some can now be discerned, but most can only be guessed at or are wholly unknown. Logically, there is every reason to believe equally drastic changes could occur in the dark side of biotechnology, that is, that devoted to nonpeaceful uses.

Terrorism and BW

So far there have been no recorded attempts by terrorist groups to use BW, although a few instances of such use by national intelligence agencies have been reported. In 1943, the U.S. Office of Strategic Services reportedly successfully used staph enterotoxin to contaminate the food eaten by one of the Nazi Germany's economists, Hjalmar Schact, to prevent his appearance at a conference.[75] After World War II, the United States began to expand its BW effort, and in 1950 a Special Operations Division (SOD) was formed at Fort Detrick with the responsibility "for developing special applications for BW agents and toxins."[76] At first its primary customer was the U.S. Special Forces but later most of SOD's work was reportedly done for the CIA under the cover name MKNAOMI.[77] Marks describes a unique arsenal of toxins and disease agents available to SOD and the CIA, including the deadly shellfish toxin carried by Francis Gary Powers on his ill-fated U-2 flight. Marks also reports that the CIA's Technical Services Staff developed appropriate delivery systems for BW agents, including a dart gun used to shoot persons with deadly, undetectable toxins.[78] It is not known if the CIA ever actually undertook any operations that involved BW usage.[79] The only known example of probable use of BW agents for assassination was described earlier in this chapter; those responsible were most probably from the Bulgarian secret police.

The use of biological agents for terrorist purposes is obviously in violation of the Geneva Protocol and the BWC. As yet no effective international mechanism is available to anticipate and counteract terrorist occurrences or to pursue perpetrators. One organization suggested to fill this void is the International Criminal Police Organization (Interpol).

Membership in Interpol is open to any country.[80] The organization serves as an international center to provide its members with technical and other information pertaining to crime. It can perform this service in two ways. First, it can call up information from its files, which contain material collected and collated over many years, on individuals, organizations, and other subjects of interest to Interpol and its members. (The files are open to all members.) Second, it can act as an intermediary between two nations, one requiring information and another possessing it. However, Interpol has no police powers whatsoever and does not perform independent investiga-

Table 7. International Measures to Combat Terrorism.

Agreement	Signed at
Convention of Offenses and Certain Other Acts Committed on Board Aircraft	Tokyo, 14 September 1963
Convention for the Suppression of Unlawful Seizure of Aircraft	The Hague, 16 December 1970
Declaration of the Strengthening of International Security	UNGA Resolution 2734 of 1970
Convention for the Suppression of Unlawful Acts Against the Safety of Civil Aviation	Montreal, 23 September 1971
Convention on the Prevention and Punishment of Crimes Against Internationally Protected Persons, Including Diplomatic Agents[a]	New York, 14 December 1973
Definition of Aggression[b]	UNGA Resolution 3314 with Annex, 14 December 1974
European Convention on the Suppression of Terrorism[c]	Strasbourg, 27 January 1977
International Convention Against the Taking of Hostages	New York, 17 December 1979

[a]This Convention (TIAS 8532) has as its aim the protection of certain public persons, including heads of state and diplomats. Provisions of the Convention come into force if, for example, there is a "violent attack upon official premises, private accommodations, and transportation" (Article 26) perpetrated in such a manner as to endanger the protected person. Thus an attack by terrorists using BW upon a capital city would inevitably endanger persons protected under the Convention, even if they were not the direct target of the attack. Therefore, Convention provisions would come into play, including Article 4, which obligates states to cooperate with one another to prevent possible terrorist acts.

[b]In Article 3 of the UNGA Resolution 3314(29) an act of aggression includes "the use of any weapons by a State against the territory of another State" (Article 36), and also "the sending by or on behalf of a State of armed bands, groups, irregulars, or mercenaries, which carry out acts of armed force against another state" (Article 3g). Under these terms a group of persons employing BW against an adversary would be committing aggression enabling the attacked State to take defensive measures it deems necessary including, perhaps, destroying the laboratories it suspects are manufacturing BW. These defensive measures could be taken pursuant to the generally recognized principles of "self-help" or "self-defense." See J. L. Brierly, *The Law of Nations*, 6th ed. (Oxford: Oxford University Press, 1963), pp. 397–432.

[c]By linking the goal of achieving greater unity between member states with the aim to still growing concern caused by terrorism, the Council of Europe drew up and adopted this Convention, the goal of which is to catch and punish perpetrators of outrages. A particularly effective means of accomplishing the goal, according to the Council, is to make certain the suspect is unable to evade justice by crossing national boundaries. The rather narrow concern of the Convention is that extradition laws of member nations conform with one another, thereby preventing suspects from taking advantage of previously existing loopholes, including (1) taking refuge in a nation that has no extradition agreements in force with the victim nation; (2) taking refuge in a nation that bars extradition of certain types of refugees, for example, those accused of political offenses; and (3) taking refuge in a nation whose extradition laws are so muddled it is possible to delay proceedings interminably.

tions. Moreover, the organization has been the target of much criticism.[81] Most often Interpol is involved in what may be termed common criminal activity,[82] and apparently, it has not been active in combating international terrorism. However, the main reason why Interpol is unsuitable for managing terrorist threats hinges on its constitution and membership. All members have right of access to the information on file, and member nations include those known to support terrorist activities. Any concerted effort against terrorists through Interpol would be compromised at an early stage.

Various international measures have been formulated to combat terrorism, usually in response to particularly noticeable terrorist acts, for example, the taking of hostages or the highjacking of aircraft. Some of the existing agreements are listed in Table 7.

Extant international antiterrorism measures lack specific instruments to control the future terrorist armed with BW weapons, and at present, the 1972 BWC provides the only specific grounds for international efforts to contain terrorists employing BW devices. A more effective approach, however, would be for the international community to attempt control over all terrorist activity. For that, additional international instruments will be needed, beginning perhaps with an international convention similar to the European Convention on the Suppression of Terrorism. If the resulting Convention was sufficiently strong, it could circumscribe terrorism generally.

Conclusion

I have argued above that most, if not all, of the technical factors, which historically may have prevented full-scale use of BW, may be modified by the emergence of rDNA research. Recombinant DNA techniques could be employed to increase controllability over potential BW agents, while at the same time stabilizing them and increasing their resistance to detrimental environmental factors. An advantage over adversaries could accrue to a nation if it decided to channel its biotechnological efforts toward the development of BW agents with enhanced disease potential and characteristics sufficiently altered to delay identification once used. A perceived advantage by a nation to wage BW could be enough of an incentive to press its claims on other nations, thereby destabilizing the international system. Nuclear weapons are still monopolized by a comparatively few nations who got together, after a fashion, to slow down nuclear spread. However, biotechnology now offers non-nuclear nations an opportunity to develop exceedingly powerful weapons at a relatively low cost.

At the same time, it should be made clear that the emergence of rDNA research and its possible BW applications does not alter the letter or spirit of the BWC. In other words, the manufacture of bacteriological and biological substances through rDNA techniques is banned unless it is for peaceful purposes. Having said this, one must unfortunately continue by pointing out weaknesses built into the Convention. Flaws include those pertaining to verification, processing of complaints, conforming national legislation, difficulties in determining peaceful versus nonpeaceful use, and new processes and materials useful to BW not included under Convention provisions.

Despite all the flaws in the BWC, under present circumstances in international politics, no better instrument can be envisioned. Unless nations are willing to forego some of the prerogatives of sovereign immunity, it is doubtful that adequate verification measures can be included within an improved BWC. A clear indication of the future of this issue may be seen by the fate of the CW treaty being negotiated between the USSR and the United States.[83] Negotiations stalled in 1980–81, reportedly because of the difficult issue of verification.[84] But if the problem is resolved, a future BWC could also benefit from lessons learned.

In a similar vein, little can be expected to occur in the international arena to anticipate terrorists employing BW agents. Attention will not be directed to this subject unless—and until—it becomes a pressing concern, which is not the case at this time.

Notes

1. N. V. Turbin, "Genetic Engineering: Reality, Perspectives, and Dangers," *Voprosy Filosofii* no. 1 (1975): 47–56; and O. Baryoan, "Dangers of Genetic Engineering Discussed," *Literaturnaya Gazeta* (Moscow), 26 February 1975, pp. 13–16.
2. United States Department of State, *Chemical Warfare in Southeast Asia and Afghanistan*, special report no. 98 (1982).
3. In the United States, the military was ordered by President Richard Nixon on 25 November 1969 to cease all BW work and begin disposing of existing BW stocks. The United States was, therefore, in compliance with the BW Convention before it came into force. See United States Congress, Senate, *Biological Testing Involving Human Subjects by the Department of Defense*, Hearings before the Subcommittee on Health and Scientific Research of the Committee on Human Resources, 1st session, 95th Congress, 8 March and 23 May 1977 (Washington, D.C.: GPO, 1977), p. 52.
4. This complex subject has been treated by F. F. Cartwright, *Disease and History* (New York: Thomas Y. Crowell, 1972), and most notably by W. H. McNeill, *Plagues and People* (New York: Anchor/Doubleday, 1976).
5. V. J. Derbes, "DeMussis and the Great Plague of 1348: A Forgotten Episode of Bacteriological Warfare," *Journal of the American Medical Association* 196 (1966): 179.
6. G. DeMussis, quoted in ibid.
7. G. DeMussis, quoted in ibid., p. 182.
8. C. G. Héden, "Defenses Against Biological Warfare," *Annual Review of Microbiology* 21 (1967): 643.
9. *M. mallei* is the causative agent of glanders which has been described as "a purulent inflammation of mucous membranes and an eruption of nodules on the skin which coalesce and break down, forming deep ulcers, which may end in necrosis of cartilage and bone." *Dorland's Illustrated Medical Dictionary*, 23rd ed. (Philadelphia: W. B. Saunders, 1959), p. 561.

10. United States Congress, Senate, *Biological Testing Involving Human Subjects*, p. 65.
11. Ibid., p. 64. See also Union of Soviet Socialist Republics (USSR), *Materials on the Trial of Former Servicemen of the Japanese Army Charged with Manufacturing and Employing Bacteriological Weapons* (Moscow: Foreign Languages Publishing House, 1950), p. 29.
12. USSR, *Trial of Former Servicemen*, p. 20.
13. Ibid., pp. 17, 289.
14. Ibid., p. 63.
15. International Scientific Commission (ISC), *Report of the International Scientific Commission for the Investigation of the Facts Concerning Bacterial Warfare in Korea and China.* Abstract reprinted in the *Chinese Medical Journal* 70 (1952): 335–660.
16. Ibid., p. 483.
17. Ibid., p. 350.
18. The 1 May 1981 BBC-TV program was discussed in B. Beckett, "N Codes for Anthrax," *New Scientist* 90 (1981): 713.
19. ISC, *Report*, pp. 478–79.
20. Ibid., p. 341.
21. Ibid., p. 399.
22. Ibid., pp. 392–98.
23. See R. A. Zilinskas, *Managing the International Consequences of Recombinant DNA Research* (Ph.D. dissertation, University of Southern California School of International Relations, 1981).
24. O. Johnston, "U.S. Links Soviet Fatalities, Germ-Warfare Materials," *Los Angeles Times*, 19 March 1980, p. 1.
25. I have written about the Sverdlovsk episode in detail in "Anthrax in Sverdlovsk: Epidemic or BW?" *Bulletin of the Atomic Scientists* 39 (June 1983): 24–27.
26. W. J. Stoessel, Jr., "Reported use of Chemical Weapons," *Department of State Bulletin* 81 (1981): 79.
27. Johnston, "U.S. Links Soviet Fatalities."
28. N. Wade, "Toxin Warfare Charges May Be Premature," *Science* 214 (1981): 34; and "Yellow Rain Riddle," *Science News*, 7 October 1981, p. 250.
29. United States Department of State, *Chemical Warfare in Southeast Asia and Afghanistan.*
30. F. Fenner and F. N. Ratcliffe, *Myxomatosis* (Cambridge: Cambridge University Press, 1965), pp. 17, 22.
31. Ibid., p. 23.
32. Ibid., p. 273.
33. Ibid., pp. 277–78.
34. Ibid., pp. 115, 277.
35. Ibid., pp. 229, 305.

36. Ibid., p. 29.
37. Ibid., p. 306.
38. ISC, *Report*, p. 362.
39. The Japanese government has recently confirmed the existence of unit number 731 and the atrocities that were committed during BW experiments. See P. Y. Chen, "Japan Confirms Germ War Testing in World War II," *Washington Post*, 8 April 1982, p. A19.
40. United Nations General Assembly, *Report of the Conference of the Eighteen-Nation Committee on Disarmament*, UN Document A/7189, September 1968, p. 19.
41. Virulence in this chapter refers to the capacity of an organism to cause high fatality rates in hosts.
42. For accounts that attempt to explain the causes for the increase in antibiotic resistance among bacteria and the costs that may accrue as a result, see B. Dixon, "Rampant Resistance," *New Scientist* 68 (1975): 194; B. J. Culliton, "Penicillin-Resistant Gonorrhea: New Strain Spreading Worldwide," *Science* 194 (1976): 1395–97; and N. K. Eskridge, "Are Antibiotics Endangered Resources?" *BioScience* 28 (1978): 249–52.
43. D. M. Glover, *Genetic Engineering—Cloning DNA* (New York: Chapman & Hall, 1980), pp. 18–25.
44. L. P. Elwell and P. L. Shipley, "Plasmid-Mediated Factors Associated with Virulence of Bacteria to Animals," *Annual Review of Microbiology* 34 (1980): 473.
45. For example, colonization antigens could in the K88 plasmid produce strains of *E. coli* enteropathic to pigs, K99 plasmids are found in *E. coli* infecting cattle and sheep, and *E. coli* carrying the CFA I and CFA II plasmids infect humans. On the general level, the presence of the Col V plasmid, which codes for the production of the protein colicin V, enhances the ability of the host *E. coli* to survive in blood, peritoneal fluid, and the intestinal tracts of man and animals. Therefore, the Col V + *E. coli* is found to cause bacteremia in both man and animal. Elwell and Shipley, "Plasmid-Mediated Factors," pp. 474–78.
46. Ibid., pp. 486–92.
47. J. Emsley and D. Pallister, "Bulgarian Brolly Baffles Germ Warfare Boffins," *New Scientist* 80 (1978): pp. 80, 92.
48. The *Washington Post* (12 February 1982) claims ricin, a derivative from the castor bean, was the responsible biotoxin.
49. For a fictional account of this possibility, see S. King, *The Stand* (New York: Doubleday, 1978).
50. The University of Maryland has a pioneering program to train biochemical engineers in rDNA techniques. The program supplements the undergraduate biology program. The head of the new program, Richard Wolf, believes most techniques can be carried out by technicians. T. H. Maugh, "First Course for Genetic Engineering Technicians," *Science* 211 (1981): 1142.
51. R. A. Zilinskas, "Recombinant DNA Research and the International System," *Southern California Law Review* 51 (1978): 1483–1501.
52. "Protocol for the Prohibition of the Use in War of Asphyxiating Poisonous or

Other Gases, and of Bacteriological Methods of Warfare," TIAS 8061 (1925).

53. N. Sims, "Biological Disarmament: Britain's New Posture," *New Scientist* 52 (1971): 19.
54. UNGA Resolution 2454A (XXIII). The Secretary-General appointed a group of fourteen experts to help him prepare the report. The terms of reference for the report were specified: "The aim of the report should be to provide a scientifically sound appraisal of the effects of chemical and bacteriological (biological) weapons and should serve to inform governments of the consequences of their possible use." United Nations, General Assembly, *Report of the Secretary-General on Chemical and Bacteriological (Biological) Weapons and the Effects of Their Possible Use*, UN Document A/7575, 1 July 1969, x.
55. Ibid., p. 116.
56. On 16 March 1975 President G. Ford, having received the advice and consent of the U.S. Senate, ratified the BW Convention. By that date sixty-one other nations had ratified or acceded to the BWC; of these nations, Austria, India, and Kuwait have attached supplementary reservations, statements, or understandings. By 1 January 1980, eighty-six nations had ratified the treaty. United States Department of State, *Treaties in Force*, Department of State pub. 9136 (Washington, D.C.: GPO, 1981), p. 268.
57. United Nations, Review Conference of the Parties to the Convention on the Prohibition of the Development, Production, and Stockpiling of Bacteriological (Biological) and Toxic Weapons and on Their Destruction, *Report*, UN Document BWC/Conf. 1/5, 8 February 1980.
58. L. Aspin, "The Verification of the SALT II Agreement," *Scientific American* 240 (1979): 38.
59. SALT II has been repudiated by the Reagan administration, but its provisions are apparently still being followed by the United States at this writing.
60. Congressional Research Service, *Chemical and Biological Warfare: Selected Issues and Developments During 1978 and January 1–June 30, 1979*, report no. 79–156 SR (1979), p. 31.
61. A bill (H.R. 8149) was introduced during the Ninety-third Congress and referred on 24 May 1977 to the Senate's Committee on the Judiciary. The bill would have provided for fines and imprisonment for any U.S. citizen found guilty of developing and producing BW; no action has been taken on this bill. Congressional Research Service, *Chemical and Biological Warfare*, p. 30.
62. In this case, exotoxin A has such low order toxicity as to be useless in war or for terrorist purposes.
63. N. Wade, "BW and Recombinant DNA," *Science* 208 (1980): 271.
64. L. R. Ember, "Yellow Rain Toxin a Potential Anticancer Drug," *Chemistry and Engineering News* 59 (1981): 29.
65. The *Washington Post* (7 February 1984) recently reported similar considerations in a military project designed to produce a vaccine against dysentery.
66. "DOD Offers Dollars for Far-Out Research Ideas in Molecular Detection," *McGraw-Hill's Biotechnology Newswatch* 1 (1981): 1.
67. These figures were compiled for the United Nations Industrial Organization, *The Establishment of an International Centre for Genetic Engineering and Bio-*

technology, UNIDO document UNIDO/IS.254, 9 November 1981, pp. 23–26. The consulting experts were H. W. Boyer, A. Bukhari, A. Chakrabarty, C. G. Hedén, S. Narang, S. Riazuddin, and R. Wu.

68. An estimate of the costs involved in starting up a nuclear research program, possibly capable of eventually producing a nuclear weapons system, can be made from Iraq's recent illfated effort. As part of an agreement France contracted to supply Iraq with a large nuclear research center; a facility to consist of two small reactors, staffed by approximately 600 people and costing around $275 million. Additionally, Italy was contracted to supply Iraq with four laboratories, including a radiochemistry laboratory for extracting plutonium from spent reactor fuel, at the cost of $50 million. The figures involved suggest the cost of staff, materials, and facilities for developing a complete nuclear weapons capability would be billions of dollars. J. Perera, "Was Iraq Really Developing a Bomb?" *New Scientist* 90 (1981): 689. (The facility was destroyed by Israeli bombing before it became operational.)
69. M. B. Callaham and K. M. Tsipis, "Biological Warfare and Recombinant DNA," *Bulletin of the Atomic Scientists* 34 (1978): 11, 50.
70. C. G. Hedén, "Defenses Against Biological Warfare," *Annual Review of Microbiology* 21 (1967): 663.
71. On 8 August 1945 an agreement was signed between the United States, Great Britain, the USSR, and the Provisional Government of France calling for "prosecution and punishment of the major war criminals of the European Axis." Annexed was the Charter of the International Military Tribunal whose function it was to try the war criminals. Appropriate to this study is Article 6(c) which defines Crimes Against Humanity as:

    namely, murder, extermination, enslavement, deportation, and *other inhumane acts committed against any civilian population before, or during the war.* . . . Leaders, organizers, instigators, and accomplices participating in the formulation or execution of a common plan or conspiracy to commit any of the foregoing crimes are responsible for all acts performed by any persons in execution of such plan. (Emphasis added.) From M. M. Whiteman, *Digest of International Law*, Department of State pub. 8354 (Washington, D.C.: GPO, 1968), pp. 882–83.

    Though the Nuremburg Principles were formulated for the purpose of judging Axis war criminals, they are now part of the body of law which attempts to set forth enduring rules for acceptable conduct in warfare. A decision-maker, familiar with the Nuremburg Principles and similar international law, would probably weigh the cost of breaching these laws when deciding whether or not to employ BW.
72. United Nations, Review Conference of the Parties to the Convention on the Prohibition of the Development, Production, and Stockpiling of Bacteriological (Biological) and Toxin Weapons and on Their Destruction, *Final Document*, UN Document BWC/Conf. 1/10, 21 March 1980.
73. United Nations, Review Conference of the Parties to the Convention on the Prohibition of the Development, Production, and Stockpiling of Bacteriological (Biological) and Toxin Weapons and on Their Destruction, *Report*, UN Document BWC/Conf. 1/5, 8 February 1980, p. 16.

74. United Nations, Review Conference, *Final Document*, 21 March 1980, p. 18.
75. United States Congress, Senate, *Biological Testing Involving Human Subjects*, p. 246.
76. Ibid., p. 245.
77. J. D. Marks, *The Search for the Manchurian Candidate: The CIA and Mind Control* (New York: Times Books, 1979), p. 74.
78. Ibid., p. 76.
79. The director of the CIA has written, ". . . we do not now support, either directly or indirectly, any research or development involving recombinant DNA but will only monitor and assess activities of others." S. Turner, letter to Paul G. Rogers, 18 July 1977.
80. Membership in Interpol is dictated by Article 4 of its constitution: "Any country may delegate as a member to the Organization any official police body whose functions come within the framework of activities of the Organization." See T. Medal-Johnsen and V. Young, *The Interpol Connection* (New York: Dial Press, 1979), p. 245.
81. Medal-Johnson and Young (*The Interpol Connection*, p. 237) offer a devastating review of Interpol's activities:

    We have shown that Interpol is guilty of selective violations of its Article 3 ["It is strictly forbidden for the organization to undertake any intervention or activities of political, military, religious, or racial character"], that it collects criminal intelligence data on individuals and groups with no criminal history, that its files contain erroneous data, that individuals [law enforcement and others] have used it to further personal goals of one kind or another.
82. Cases handled by Interpol's U.S. representative involve, among other things, transnational movements of persons suspected of criminal activity in the United States (robbery, fraud, various narcotic offenses, and so forth), international movement of narcotics, illegal manufacture and use of documents (for example, passports), and similar cases.
83. See M. S. Meselson and J. P. Robinson, "Chemical Warfare and Chemical Disarmament," *Scientific American*, 242 (1980): 38–47, for a thorough review on the state of CW as well as the peace/war issues that surround the field.
84. J. C. Burton III, "CB Winds of Change," *Defense and Foreign Affairs* 7 (1980): 32; and L. R. Ember, "Scientific Gaps Cloud New Chemical Arms Issue," *Chem. & Eng. News*, 59 (1981): 21–24.

Added Note: Several articles have recently appeared on the subject of biotechnology and BW: S. Wright and R. L. Sinsheimer, "Recombinant DNA and Biological Warfare," *Bulletin of the Atomic Scientists* 39 (1983): 20–26 and E. Geissler, "Implications of Genetic Engineering for Chemical and Biological Warfare" in *World Armaments and Disarmament: SIPRI Yearbook 1984*, (London and Philadelphia: Taylor and Francis, 1984), pp. 421–51. For a more complete account of the anthrax bomb episode (note 18), see R. Harris and J. Paxman, *A Higher Form of Killing: The Secret Story of Chemical and Biological Warfare*, (New York: Hill and Wang, 1982), pp. 100–6. For a Soviet rejoinder to the charge of "yellow rain" (note 26), see United Nations, "Letter Dated 20 May 1982 from the Permanent Representative of the USSR to the United Nations Addressed to the Secretary General" and Annex; document A/37/233, 21 May 1982.

What does it all mean? Where will the extraordinary gains in knowledge and the development of powerful techniques for analyzing and manipulating biological molecules take us? For the scientist, it means learning a great deal more than we used to know. But for most people, it means new technology. What can science do for me? That is the question posed in the funding committees of Congress, in the venture capital organizations, and in the new biotechnology companies. And the ordinary citizen, who values the conveniences of modern technological life and freedom from disease and poverty, perhaps wants to know more than anyone.

Zsolt Harsanyi is a geneticist who came to Washington to become project director of an Office of Technology Assessment study published in 1981 on the emerging applications of gene splicing in all areas except for direct human genetic intervention. Now a director of Porton International Ltd. and an advisor to E.F. Hutton, Dr. Harsanyi has been involved in making the dreams of the 1970s come true in the 1980s, establishing commercial ventures in biotechnology. He asks us to look ahead twenty years and attempts to give us a glimpse of what lies beyond today. But in the time scale of modern scientific discovery, twenty years is a very long time. Some may view his predictions as fantasy; others will see them as shortsighted. As he himself admits, predicting the future is a dangerous exercise in that the farther in the future one looks, the more likely one is to be wrong. In any case, it is universally acknowledged that the changes in technology and in society that will eventually result from genetic manipulation will be extraordinary.

11

Beyond Recombinant DNA: The Next Twenty Years

ZSOLT HARSANYI

During the last fifteen years, some very esoteric basic research has ushered in a new age for industry and medicine. Hundreds of articles have hailed the advent of the Age of Genetics, or variously, the Age of Biology. In fact, the last few years have witnessed in the news media not only a flood of old words incorporating the prefix "bio," but also a host of neologisms: bioengineering, biotechnology, bioresources, bioindustries, bioconversion, and so on. The latest edition of the comprehensive Oxford University dictionary fails to include over two dozen of the commonly used terms incorporating the prefix bio. In addition, the novelty of these words is underscored by the recent decision by the United States Patent and Trademark Office to grant a trademark for the word biotechnology in the title of a publication.

Why is there all this current interest in biology? It is largely because the science has been surrounded by a plethora of promises and possible perils, all of which were based on speculation. In the mid-1970s, society found itself with some powerful new biological tools, but it didn't really know their potentials or their limits.

Consider the promises. Headlines read: "Interferon—the new wonder drug against cancer!" Genetic engineering and biotechnology have been hailed as a panacea for all the world's problems. These can be summarized as the five "F's": food and feed, fuel, fiber, and if one pardons the poetic license, pharmaceuticals.

Consider the perils. Headlines read: "Oil-eating bacteria might endanger our fuel supplies!" The bioengineers have been accused of disrupting not only the population's health and environment with monster mutants, but also the very moral fabric of society. But both sides have been characterized by hyperbole. Most proponents and opponents see apparently unlimited possibilities—or, more accurately, uncertainty—about the limits of the possibilities. And this is where the hope lies and where the fear lies.

There is one big difference, however. Even early in 1980 the critics of recombinant DNA argued that the promises of the technology were no less hypothetical and speculative than the risks they were accused of overstating. But by mid-1981 it became clear that the promises were becoming more of a reality. Human insulin produced by gene cloning had entered clinical trials. The successful production of foot-and-mouth disease vaccine material was announced in June 1981. And approvals for the large-scale manufacture of a half dozen other rDNA products were being obtained.

The risks of cataclysmic disaster, on the other hand, have remained speculative; there is still no demonstrable hazard associated with random gene manipulation.

The excitement surrounding biology was the result of two trends that emerged and then merged in the past decade to focus interest on the industrial applications of the science. First, there was the creation of powerful new tools useful in production; the cumbersome biological sledgehammers could be replaced with exquisitely fine scalpels. And second, there was the recognition of society's dwindling raw material and energy base. It is becoming clear that there will be an increasing need to depend on biological processes and systems to provide material needs. The first trend, then, emphasizes the role of biology as a tool; the second, as a raw material. Continuation of both trends through the next twenty years—almost a certainty—will assure a firm role for biotechnology in meeting many of society's needs.

The availability of new tools is largely, but not exclusively, due to significant advances in genetics. The first Office of Technology Assessment report, which was published in April 1981, concluded that biological processes, which had been merely laboratory curiosities, can play a significant role in large-scale industrial technologies.[1] Highly versatile laboratory manipulations—rDNA or gene splicing, cell fusion or hybridization, and others—can all be used to alter the heritable characteristics of cells in making products. The alterations can either increase production of existing products or can generate completely new products.

Although the initial excitement was over the new laboratory techniques of molecular genetics, it soon became evident that an altered gene does not immediately translate into a new product. The altered genes, or a totally new combination of genes, must play a role in some industrial system of production. That system can be through plants and animals in agriculture or by other means such as bacteria in fermentation, but by and large the production system is a biological one.

Recognition of this relationship between genetics and biological methods should clarify much of the confusion surrounding the terminology of genetic engineering. Unfortunately, the terms biotechnology, genetic engineering, and recombinant DNA are often used interchangeably. In fact, biotechnology refers to the use of living microorganisms, animals, and plant cells, or their components such as enzymes or membranes, to transform, synthesize, degrade, or concentrate materials. Therefore, it is a method of production or a process used by a variety of industries, rather than a class of products produced by a single industry. It is not analagous to industries recognized by their product, such as the automobile, textile, or computer industries. It would be more appropriate to use the plural, biotechnology industries, in reference to those industries that make use of biological techniques.

In contrast, genetic engineering is not so much an industry as a laboratory-scale operation in which the hereditary material of a cell is altered. The term has come to be applied to a variety of molecular and cellular techniques with names such as cell fusion or hybridization, transformation, chromosome-mediated gene transfer, and rDNA.

It should be clear then that rDNA or gene splicing refers to just one, albeit the most powerful, method of altering the hereditary characteristics of a cell. It is not synonymous with genetic engineering.

What genetic engineering does is to alter a cell so that it can produce more or different chemicals or to perform better or new functions. The altered cell, or more appropriately the population of altered cells is, in turn, used in an industrial process.

Practical application of genetics, of course, has been known for over a century. Historically, the most significant impacts have been in the field of agriculture. The development of hybrid corn in the 1930s, and later the Mexican dwarf wheat and IR-8 variety of rice have profoundly affected society's ability to feed billions of people. Classical breeding, the progenitor to direct genetic engineering of the hereditary material, was responsible for the 115 percent increases in wheat yield, 320 percent increases in corn, and 413 percent increases for processed tomatoes over the years 1930 to 1975.[2] Genetic manipulation of bacterial strains has also been responsible for the several hundredfold increases in the industrial production of antibiotics and amino acids. It has been almost automatic that if a microorganism is found to produce a useful product, genetic strain improvement will be used to maximize output.

Nor is biotechnology really neotechnology, for fermentation has been used by man for centuries in food and beverage production. Currently, over 200 products are produced in bulk quantities by biotechnology or fermentation methods[3] (Table 1). The significance of this fact lies in biotechnology's historically proven success.

Table 1. Commercial Fermentation Products: 1982.

Product class	Number of different products in class
Antibiotics	90
Enzymes	40
Amino acids	20
Organic acids, solvents	20
Vitamins, growth factors, etc.	10
Miscellaneous (flavors, insecticides, etc.)	25
Total	205

The advent of rDNA and other gene-transfer technologies has opened up the possibility of biological transformations never dreamed of before. These new tools could lead to new products or alternative production methods by way of two avenues. First, genes could be transferred between organisms which would normally have natural barriers to such genetic mixing. As a result, new properties could be conferred on the recipients of foreign genes. Not only have primitive bacteria been endowed with traits from other primitive bacteria with which they normally would not mate, but even higher mammals such as mice have been the recipients of genes that coded for alien proteins. In one such case the new trait was the heritable production of rabbit hemoglobin.[4] Secondly, genetic engineering could increase the production of useful substances in a variety of cells. In hybridoma technology, genetic engineering led to the stable formation of cell lines that possess the unlimited lifespan of one parental cell and the ability to produce homogeneous antibody preparation of the other parental cell. The net result is a continued supply of specific antibodies. Recombinant DNA too could yield large quantities of desired substances, simply by transferring multiple copies of a gene into a recipient cell. Every time the cell divided, a commensurate number of gene copies would be replicated as well. If the gene is to be viewed as the new transistor, as some have claimed, cells used in biotechnology production methods would have to be viewed as the new amplifier.

In summary, the genetic technologies are making available a host of new diagnostic and therapeutic substances in quantities previously unavailable. As a result, these technologies are responsible for a paradigm shift that will have long-range implications for industrial production. Biology, like chemistry and physics, has now been recognized as being able to provide effective tools for solving problems. A striking example is pollution control. Directors of research at the U.S. Environmental Protection Agency have explained that most of their engineers have thought in terms of classical engineering solutions: draining, physical containment, chemical disinfection, and so forth.[5] Only recently has their attention been turned seriously to biological solutions, such as genetically modified pollutant-degrading microorganisms.[6] As a result, over the next twenty years they will increasingly explore innovative biological methods of pollution control. Similarly, each industry is in the process of recognizing biology's potential role in solving its problems.

Each of the industrial needs listed in Table 2 can in selected cases be met with a biological solution. The biological approach will be used in either of two ways: to improve the production method for a substance already in production or to produce a substance hitherto unavailable in significant amounts. The needs for alternative raw materials, decreased energy requirements, and less toxic by-products all fall within the category of production improvements. In some cases, an improvement means replacement of a currently acceptable technology with a biotechnological approach. At the same time, increased production of scarce materials and novel products or processes are continuing needs for all industries.

Table 2. Major Industrial Needs by Year 2000.

1. Alternative raw materials
2. Decreased energy requirements
3. Less toxic by-products
4. Increased production of scarce materials
5. Novel products or processes

Problems such as the need for alternative raw materials, decreased production energy, and novel products can be expected to span a variety of industries, as shown in Table 3. Biotechnology is accelerating developments in the different industries because it has awakened interest in biology as a technology. It has encouraged companies to ask, Where can biology improve or replace my product or process? And it has encouraged scientists to ask, Where can I apply my findings? Much of biology has a wealth of applications, but it takes a new kind of thinking, a new commitment, to put the basic knowledge to use.

Because of its versatility, the technology will have an impact across the entire spectrum of chemical substances. Theoretically, virtually all organic

Table 3. Industries Expected to be Most Affected by Biotechnology.

1. Drugs and pharmaceuticals
2. Health care
3. Bulk chemicals
4. Specialty chemicals
5. Food and agriculture
6. Toxic waste and pollution
7. Mining
8. Energy
9. Other manufacturing

chemicals—plastics, flavor, perfume materials, synthetic rubber, medicinal chemicals, pesticides, industrial solvents, and so forth—are candidates for production by biological methods. Nevertheless, the specific products that will be affected in each chemical group can only be identified on a case-by-case basis. For each chemical, the relative merits of competing technologies will determine where the economic threshold will lie for adopting a biotechnological process. Competition will be found between different biotechnological methods, as well as between nonbiological methods. It will not be enough to know that a biotechnological method will be used to produce a substance, it will be important to know which one. As shown in Table 4, at least five types of technologies can be expected to compete in the future.

For example, insulin can be produced by several competing processes. Chemical synthesis was shown to be possible twenty years ago, but it is still prohibitively expensive on a commercial scale. Extraction of insulin from pancreatic tissue is the current method of choice. Fortunately, animal insulin—either bovine or porcine—is sufficiently similar to human insulin to allow its use in therapy by 95 percent of diabetics. It is also available in large quantity. Clearly, if the patient population demanded insulin extracted from humans, the supply would be extremely difficult to provide with current production methods. Other methods produce insulin in tissue culture, that is, by pancreatic cells grown in large glass vessels filled with nutrient fluids.[7] The quantities of insulin obtainable by this method, however, are still limited. Production by microorganisms carrying the human insulin gene in rDNA form offers practically unlimited quantities. This type of competition among several methods of production will be characteristic for most chemical and pharmaceutical compounds. It is safe to say that feasibility studies are being carried out for the production of all the major chemicals to determine whether biotechnology will play a dominant role. In some cases, however, combinations of methods will be employed, particularly as mixed biological/chemical syntheses. Novo Industrials, for example, follows a two-stage process in which porcine insulin is extracted and then enzymatically altered to resemble human insulin.[8]

But the above comparison would be incomplete if another realistic alternative isn't mentioned. For many diseases such as diabetes, therapy might be achieved by transplanting the appropriate tissue into the patient or ultimately by transferring the normal hormone-producing gene apparatus into the cells of the patient. But the significant point is that no single method will be best for all substances.

Some people are highly optimistic about biotechnology's eventual competitive edge. The president of one of the first genetic engineering companies has said, "Anything organic chemistry can do, biotechnology will do better."[9] And the vice president of a leading pharmaceutical company opined, "The potential for recombinant DNA is bounded only by one's imagination."[10] However, these statements are truer in theory than they will ever be in practice, as the example below demonstrates.

A corporate decision to use a biological production method would be made largely on the assessment of competing production costs. This must be done on an individual basis for each of the thousands of substances potentially amenable to the new methods. And the relative costs can change significantly over time and depending on geography.

Table 4. Competing Technologies for Producing Chemical Products.

1. Chemical synthesis
2. Extraction from natural sources
3. Production in cell (tissue) culture
4. Fermentation by genetically engineered microbes (e.g., recombinant DNA)
5. Mixed biological/chemical synthesis (semisynthesis)

For example, in the late nineteenth century, Louis Pasteur discovered that some bacteria can produce acetone and butanol; they were able to convert raw materials rich in carbohydrates to these organic solvents. The finding remained a curiosity until the early twentieth century when military demands during World War I caused a shortage in these materials. Acetone was needed for the manufacture of cordite, a naval explosive; and butanol was needed for the production of synthetic rubber. As a result, microbiologists began intensive work to select genetically superior strains and to develop advanced fermentation methods for converting inexpensive raw materials such as corn starch, molasses, potato mash, or sugar cane to the organic chemicals. Plants were constructed in Great Britain, Canada, the

United States, and India to take advantage of the new developments.

However, their operation did not continue for long. Soon, petroleum and natural gas became increasingly less expensive as an alternative raw material for the production of organic chemicals. Chemical synthetic methods replaced biological synthetic methods. Biology's bid for an industrial role was merely a flash in the pan. To be sure, some countries continued to use fermentation methods for production of certain bulk organic chemicals. South Africa and the Soviet Union, for example, still find that fermentation is the method of choice in the production of acetone, butanol, and ethyl alcohol. And the reason is simple. No matter how you clothe it, whatever euphemisms you might use, the bottom line is economics.

In the early 1970s, the economic variable began to change abruptly. As the cost of the fossil-fuel based raw material spiraled upwards, the accepted methods of hydrocarbon chemistry began to look less attractive. The industry hasn't seen the intersection of the two cost curves yet, where the biological production costs drop below the chemical ones. But that trend could be accelerated either by even further increases in petroleum prices (after the current dip) or by engineering bacteria to utilize the potentially most inexpensive and abundant of organic raw materials: cellulose from wood.

Whether for bulk organic chemicals or pharmaceuticals, a recognition of the relative merits of biotechnological production methods will be useful in identifying the sequence of impacts on various products from now to the year 2000. These merits will vary from substance to substance, but certain generalizations can be made. The promise of genetic engineering to improve biotechnology's competitive edge can be examined in four categories of substances: large polypeptides or proteins, peptides or small proteins, nonproteinaceous organic chemicals, and entire functioning cells.

At present, biotechnology shows greatest promise in those areas in which alternative technologies offer the least competition, the production of large proteins. Current chemical synthetic methods are simply impractical for proteins larger than thirty to forty amino acids. Hence such large molecules must either be extracted from natural tissue sources or produced by microorganisms (cells) carrying the appropriate gene. Recombinant DNA will also have its initial impact on these substances because of the nature of the technology. In particular, the technique is most useful when a single gene that codes for the product can be identified. Once identified, it can be transferred, cloned, and expressed in the microorganism of choice. Such genes have been identified mostly for protein products, although exceptions can be found for the production of smaller organic molecules like amino acids and antibiotics. For proteins, then, unless some as yet unforeseen advances in chemical synthesis are developed, biotechnology will be the production method of choice through the next few decades.

Most of these large proteins are of interest to the pharmaceutical industry. The technology offers proteins in larger quantities and higher qualities than previously possible. This will be of value for diagnostics that require highly specific reagents. As a variety of enzymes becomes more available, new therapeutic advances can be expected. For example, the embolism-dissolving enzyme urokinase, which has been used for some years in Europe and Japan, is increasingly gaining acceptance in the United States. Antibodies will be used more frequently in therapy, especially for cancer. By attaching cell-destroying drugs to antibodies specific to cancer cells, the drugs will be carried selectively to the target cells. Vaccines will be prepared against protozoal parasites and viruses that can be grown in sufficient quantities to make commercial production practical. The impact will be on such intractable diseases as malaria, river blindness, and hepatitis, which afflict hundreds of millions of people around the world annually.

Already one U.S. company has succeeded in using rDNA to produce high levels of hepatitis B antigen in a mammalian tissue culture system.[11] The antigen can be potentially used as a vaccine or as part of a test kit for detecting carriers of the virus among blood donors. Heretofore, the preparation of such kits has always posed a risk of infection; with the new method, no infectious virus is ever needed. One only uses the piece of the viral gene that codes for the antigenic material.

In theory, with the advent of gene transfer technologies, the gene coding for any useful protein can be spliced into a suitable microbe. This will lead to the production of many new substances that were difficult to obtain previously. But there is a major caveat: The new technologies can promise new uses for these products. For example, of the approximately fifty human protein hormones that could be made readily by rDNA technology, only ten now have proven or accepted uses. In addition to these proteins, there are hundreds of other known, or yet to be discovered, human substances that might be medically useful. Nobody know which ones, however. Nobody has ever had enough of these compounds to test their clinical usefulness. It's a chicken-or-egg problem: scientists have not developed technologies to produce enough material because they didn't know whether they'd be useful. And they didn't know whether they'd be useful because they never had enough material to test. The new technologies will offer society an opportunity to escape this vicious circle.

One of the big problems facing industry and the government research agencies is which substance to test. Which ones should be given priority? There are dozens of reported compounds that affect the immune system alone, and the costs for producing and testing these compounds will require federal cooperation (see Chapter 12). Currently, the Biological Response Modifiers Program at the National Cancer Institute is actively pursuing research on molecules that regulate cell activity. For fiscal year 1980, $31.2 million was allocated for this area, of which $11.02 million was for interferon.[12] At this rate of funding, it can be expected that many substances will remain untested through the year 2000.

Estimates of the technologies' economic impact will necessarily depend on the usefulness of the product. They range from a few million dollars per year for clinically accepted drugs like insulin, growth hormone, and serum albumin to several billion dollars if one or more of the potential antiviral or anticancer agents prove to be useful. The limitation for most proteins is no longer the production technology, but the pharmacological and clinic research needed to demonstrate efficacy.

The second category of compounds, the small peptides, faces similar uncertainties. The promise of the peptide MSH/ACTH 4-10 is that of enhancing memory and concentration,[13] of bombesin to suppress appetite,[14] and of prolactin to act as a male/female contraceptive.[15] The production of compounds such as these by biotechnology will meet strong competition from chemical synthesis. As I heard one scientist comment at a meeting on biotechnology, "We chemists aren't sitting idly by while the biologists move inexorably ahead." However, we expect that over the next twenty years the learning curve for biotechnology will help advance its advantage. A single cell, for example, might be engineered to have several hundred genes for the small peptide rather than just one. Nevertheless, in some cases it will be desirable to produce peptide analogs using chemical transformations which biology cannot provide. These analogs might retain the desired biological activity without undesirable side effects.

As mentioned earlier, the availability of new technologies has been only one reason why biotechnology production methods are attracting so much attention. The 1970s brought attention to both the dwindling raw material base and the increasing cost of fuel necessary for production. Hence, two features of biotechnology stand out as highly desirable: (1) its ability to use renewable biomass resources as raw material, and (2) its use of physiological temperatures and pressures, which lead, in many cases, to reduced energy consumption.

But the future of biotechnology in producing organic chemicals will depend on several factors, each of which is difficult to predict. For example, it is imperative to find new raw materials to replace fossil fuels. This is particularly true for the bulk chemicals such as ethanol. Large amounts of raw material will be needed if alcohol continues to be used as a fuel (gasohol) or becomes the source for other organic materials. For example, ethanol can be converted to ethylene, which serves as the starting material for about 50 percent of all our organic chemicals. Brazil has already begun to design its chemical industry to use ethylene derived from biomass, particularly sugar cane.[16] Elsewhere, it will be imperative to find an inexpensive raw material that is available in bulk quantities, such as lignocellulose.

Current biotechnology methods of ethanol production use corn starch in the United States, sugar cane and cassava in South America, and sugar beets in Europe. However, only cellulose, available from forests, urban wastes, and such novel sources as kelp, meets the criteria of potential high-volume and low cost.

The third category, nonprotein organic chemicals, is by far the largest and most varied. It includes not only the 5000 or more relatively simple organic chemicals, but also gases such as hydrogen and more complex compounds such as vitamins and antibiotics.

Antibiotics are already being made by biotechnology, but the advent of the new technologies has increased the possibilities for greater variety and quantity. Of the thousands of antibiotics detected to date, hundreds have not been pursued because the microbes could not produce sufficient amounts. The new tools of rDNA and cell fusion should allow novel approaches toward increasing production. For example, the techniques could increase the number of copies of a gene that codes for a rate-limiting enzyme. As a result, in the next twenty years those antibiotics that were previously deemed uneconomic to pursue will be a source of new products.

In the pharmaceutical industry most of the medicinal chemicals are currently synthesized chemically or are extracted from living matter. If medicinal chemicals are to be made by biotechnology, enzymes must be found to carry out the reactions. Acetaminophen, for example, the largest selling, nonaspirin analgesic (the active ingredient in Tylenol), is currently made by chemical methods. In a feasibility study carried out by Genex, Inc., a chemical raw material, aniline, could be converted through two steps to the final product.[17] These conversion steps could be performed by several known fungi. According to Genex calculations, the biotechnology production could be carried out at a savings of $.55 per pound. This would translate into an annual cost savings of $100 to $200 million to the producers.[18]

Similar feasibility studies will be carried out by most chemical companies over the next twenty years. The general approach for the production of each chemical will be analogous to that used for acetaminophen: to find an enzyme or set of enzymes, either within cells or extracted from them, that can convert some raw material to the desired product. The present limitation is the lack of an economically favorable technology to convert potentially inexpensive raw material cellulose to its component sugars. It is these sugars that can then be converted to ethanol. The timing for the breakthrough cannot be predicted, but most experts are convinced that it will be achieved within a decade.

For other chemicals—the lower-volume specialty chemicals—biotechnology can be expected to continue to offer possible alternative routes for synthesis. As more traditional chemists recognize biology as a tool, mixed chemical-biological synthetic methods will be used. Chemical reactions will be employed for some steps in a multistep production method, while biological reactions will be used for other steps. The availability of enzymes needed to carry out the biological reactions will encourage such production methods. It is, in fact, rDNA technology that will accelerate this development. An increase of up to 5000-fold (above the levels found in wild-type cells) have been already achieved for several proteins in the bacterium *E. coli*.[19] A substantial cost of the Cetus process for making ethylene glycol by biotechnology is in the supply of enzymes needed.[20] But microorganisms that supply the enzymes can be genetically manipulated to produce large quantities.

Nevertheless, making a better microbe or producing more enzymes is only part of the solution. A second limitation lies in the engineering needed for scale-up. The Cetus ethylene glycol-fructose process, for example, worked well on paper. It worked well in the lab. But it reportedly faced difficulties in scale-up.[21] Moving from a laboratory process to an industrial process involves more than just using more reagents in bigger vessels. With increased scale, problems arise in the diffusion of nutrients or raw materials to the cells or enzymes, in cooling of the fermentation broth, and in a host of other activities, such as in maintaining contamination-free conditions. Solutions for these problems often depend on engineering, not biology. A mechanical system, for example, might have to be devised to rid a large reaction vessel of a toxic by-product rapidly enough to prevent inactivation of the whole fermentation. In small vessels the by-products might be removed easily, but in a large vessel, an elaborate network of pipes might have to be designed to achieve the result. In short, the ultimate success of biology as a technology will require continued advances in large-scale fermentation and biochemical engineering. Many of these advances can be expected to occur in Japan because of the availability of trained manpower. The United States is beginning to recognize its need for more engineers in this area and should establish appropriate training and research programs.

Where living cells themselves are the products, the major advances will appear more slowly. These products fall into two groups. In one group microorganisms will be developed for intentional release and use in the environment; in the other, they will be used as a source of protein for human or animal consumption.

In the first group, microorganisms are being designed to leach metals out of low-grade ore, to produce substances in oil reservoirs to enhance oil recovery, or to degrade toxic chemicals and other pollutants. It will be difficult to engineer organisms for these tasks because so little is known about their genetics and physiology. Their use in the open environment, in addition, will be hindered by federal regulatory pressures. Nevertheless, the sheer magnitude of the potential markets will make certain that both government and industry will develop the technology. According to the Environmental Protection Agency, nearly 50 million metric tons of hazardous waste are generated annually, a large portion of which are organic chemicals.[22] And approximately $2.5 billion are spent on pollution control by chemical-related industries.

Similarly, the targets for enhanced oil recovery are also extraordinarily large. Over half the petroleum remains in most reservoirs after recovery by primary methods are exhausted. Both the U.S. Department of Energy and oil companies have demonstrated interest in using microbes for recovery. Microorganisms will be marketed, possibly by companies yet to be established, for injection into oil reservoirs. Their function will be to survive long enough to produce chemicals and gases that can increase the flow of the oil. With escalating oil prices, it is expected that within twenty years the cost of producing and using the organisms in large quantities will be readily offset by increased oil recovery rates.

Finally, one more example of using live microorganisms is in the accumulation of such metals as copper, gold, uranium, silver, zinc, and vanadium, and rare earths. Because bacteria, yeasts, and algae are often capable of bioaccumulating hundreds to thousand times higher concentrations of metals than the water they are placed in, they offer a method for scavenging precious metals from mine tailings or the ocean.

Microorganisms are also used in the production of food or animal feed. This has been encouraged by some rather grim projections of the future. The *Global 2000 Report* to the President of the United States projected "that population would increase from 4 billion to 6.35 billion . . . that the real cost would rise everywhere . . . and that productive grasslands and croplands will turn to desert-like conditions."[24]

As a result, the interest in augmenting the world's supply of food has focused attention on microbial sources of protein as food for both animals and humans. Since a large portion of each bacterial or yeast cell consists of proteins (up to 72 percent for some protein-rich cells),[25] large numbers have been grown to supply single-cell protein (SCP) for consumption. The protein can be consumed directly as part of the cell itself or can be extracted and processed into fibers or meatlike items. Advanced food processing technologies can combine this protein with meat flavoring and other substances to produce nutritious food that looks, feels, and tastes like meat.[26]

To date, the major limitation on commercial production in the United States has been the cost of competing sources of protein, primarily soybeans. In other parts of the world—for example, Europe and Asia—the production of SCP was over two million tons in 1981.[27] In addition, as an example of the potential significance of SCP, the Soviet Union, which is one of the largest producers, expects to produce enough fodder yeast from internally available raw materials to be self-sufficient in animal protein foodstuffs by 1990.

Integrated systems can be designed to couple the production of a product or food with SCP production from waste. The waste sawdust from the lumber industry, for example, could become a source of cellulose for microorganisms.

Imperial Chemical Industry's successful genetic engineering of a microorganism to increase the usefulness of one raw material (methanol) should encourage similar attempts for other raw materials.[29] Hence, the transformation of inedible biomass into a source of animal feed or human food remains a major reachable target.

Before concluding this discussion of the future of biotechnology and genetic engineering, it would be appropriate to make reference once again to society's paradigm shift. Living cells and their products will offer more than just direct products or processes. They will also offer models upon which to base physical processes. This has been referred to as "biomimetics" or "biomimicry." Two examples are already available: the generation of hydrogen and the development of microcircuits.

Hydrogen is one of the great prospects for energy sources in the next century. It was discovered about eight years ago that light can split water into hydrogen and oxygen in the presence of artificial membranes containing chlorophyll.[30] The goal is to harness solar energy to generate large quantities of hydrogen from an abundant raw material, sea water. Today, artificial systems can only produce hydrogen for a few hours before breaking down, but as one learns more about the biological processes at work, one can expect to mimic them with systems that last years. For example, inorganic catalysts such as platinum dioxide might replace the unstable hydrogenase enzyme.

The development of "biological microcircuits" is even more intriguing. Gentronix Laboratories, Inc., in Maryland, is actively pursuing the construction of microcircuits that exploit the conducting and semiconducting properties of proteins. The new devices could produce a new generation of microdevices that are not only smaller than today's microelectric components but are biocompatible. This latter property would allow the electronic devices to be incorporated into living systems. One immediate application could be in reactivating the optic nerves in blind people.

All this might seem like a panacea for the world's economy. By manipulating the right genes in the right biological systems, one can come up with anything. This is the view of the gene as an genie. Open up the bottle and the genie grants one's wishes.

But some people argue that what is really being opened is a Pandora's box, which will release plagues on all mankind. That fear has been most strongly associated with rDNA. Perhaps it shouldn't be all that surprising—rDNA is intimately associated with the term "mutant." What thoughts come into people's minds when they hear "mutant"? Are they "Andromeda strain," Frankenstein's Monster, the creature from the Black Lagoon? Or are they improved dairy cows, improved wheat and corn, better penicillin-producing fungi? Most likely it's the former group of images.

As a result, controversy has surrounded the technology from its inception. The major controversy has centered around unforeseen risks, for example, microorganisms thought to be safe but which prove to be harmful. Discussion of this kind of risk is made difficult by one's inability not only to quantify the probability of occurrence but also to predict the type of harm that might occur. Enumerating the different types of damage is limited only by the imaginations of those who build scenarios of disaster. They have included epidemics of cancer, spread of oil-eating bacteria to oil tanks, uncontrolled spread of new plant life, and infection with hormone-producing bacteria. They are particularly frightening because the scenarios include not only everything people can imagine but those they cannot. This is the don't-fool-with-mother-nature philosophy.

In the controversy surrounding genetic engineering, problems have arisen from our inability to distinguish the possible from the probable. As an analogy, it is, for example, possible that one will be killed by a meteor falling to the ground, but it isn't probable.

The first step in estimating risk is identifying the potential harm. Since no dangerous accidents are known to have occurred, the types of potential harm remain conjectural. Identifying potential harm rests on intuition and on arguments based on analogy. That is why experts disagree. There is no uncontestable scientific method that dictates which analogy is useful or acceptable. By their nature, all analogies share some characteristics with the event under consideration, but differ in others.

As one such analogy, new strains of influenza virus arise regularly by natural recombination. Epidemics then result because populations have never been exposed to them and therefore have no protective antibodies. Yet is this analogy sufficient to suggest that weakened strains of *E. coli* such as K-12 could be transformed into an epidemic pathogen? Probably not.

Data gathered to date generally support the opinion that scenarios of doom or catastrophe are highly unlikely. This is the general concensus of specialists, not only in molecular biology but in population genetics, microbiology, infectious diseases, epidemiology, and public health. The scientists who convened under NIH support at Falmouth, Massachusetts in 1978 to assess these issues confirmed this view.[31]

Experts in infectious diseases have stressed repeatedly that the ability of a microbe to cause disease depends on a multitude of factors, all working in a highly coordinated manner. Supporting evidence is provided by experiments in which the most commonly used bacterium, *E. coli* K-12, could not be converted to a harmful strain even after known virulence factors were transferred to it.[32] Hence, the accidental construction of a new epidemic strain is unlikely.

Still, the possibility of creating a dangerous organism does exist. Remember, society is dealing with a double-edged sword: the same unlimited possibilities that might bring them better living through biology might also bring them a microbe monster. Hence, an important principle emerges from the debate over safety. *Society must decide* whether the burden of proof rests with those who demand evidence of safety or with those who demand evidence of hazard. The former group would halt experiments until the experiments are proved safe. The latter group would continue experiments until it is shown that the experiments might cause harm.

A significant theoretical difference exists between the two approaches. Evidence can almost always be provided to show that something causes harm. It can be demonstrated, for example, that polio virus causes paralysis, that a *Pneumococcus* causes pneumonia, and that a rhinovirus causes the common cold. It cannot be demonstrated that rDNA molecules can never cause damage. It can only demonstrated that these events are unlikely. Hence, society must determine what level of uncertainty it is willing to accept. (For a further discussion of this issue, see Chapter 9.)

As a result of decreased perception of risk and increased perception of promise, the speed of developments in the genetic-engineering industry is truly phenomenal. The number of publications on rDNA, as recorded in the Medline index, has doubled every single year since 1974. The number of companies established solely to commercialize genetic engineering is now well over 100,[33] of which over half are devoted almost exclusively to cell hybridization. And perhaps a total of 200 companies are using the techniques of rDNA. In the United States, research budgets are also increasing with the entry of the private sector. According to calculations made by Nelson Schneider at E.F. Hutton & Co., five years ago the federal government outspent the private sector in biomedical research by 19 to 1. In fiscal year 1982 the National Institutes of Health spent more than $2.4 billion, while private companies spent $400–500 million or a ratio of 5 to 1. This ratio will soon approach 1:1. In fact, if one looks at rDNA research alone, industry already outspends NIH's annual $90–100 million research budget.

With this infusion of industrial support, even greater strides can be expected than those of the past few years. And they have been substantial. At the first E.F. Hutton conference on biotechnology and genetic engineering, held in September 1979, the speakers, who included the presidents of three major genetic engineering companies, were asked when the interferon gene would be cloned. They all said daringly that it would be done in two years. Less than four months later, rumors that the gene had been cloned were confirmed. Equally phenomenal was the commercialization of human insulin made by rDNA. The time span from the first experimental cloning of rat insulin gene to the cloning of human insulin gene, its large-scale production, and clinical testing in human beings was four years. And again, it was only a year between the granting of government approval to clone the gene needed for the production of foot-and-mouth disease vaccine and its accomplishment. Finally, even the application of the techniques of genetic engineering to human beings has moved faster than anyone expected. One Nobel laureate stated a few years ago that molecular gene transfer might be attempted in man in as little as five years.[34] One year later, while about a hundred scientists were analyzing the potential hazards of rDNA at a conference in Pasadena, the announcement was made that Dr. Martin Cline at the neighboring University of California, Los Angeles, had attempted to transplant genetically altered bone marrow cells into two patients, one in Italy and one in Israel.[35] It should not go unmentioned that this feat was far from the genetic engineering of human beings that some people feared. There was no attempt to alter the entire genetic makeup of the individual. Only a single tissue, the bone marrow, was to be altered. In that sense, it would have been analogous to an organ transplant, where a defective organ is simply replaced by a healthy, functioning one.

Nevertheless, the pace of development has been surprisingly rapid, and it is tempting to speculate where it will all lead. In 1978, when members of Congress requested a study from the Office of Technology Assessment on the potential impacts of the technology, they were interested in knowing how our way of life would be altered. What would be the direct impacts—sociological, ecological, economic, etc.? The indirect impacts?

It soon became clear during the course of the study that the impacts would be not only profound but broad. Beyond this somewhat vague generalization, it would be difficult to provide a concise summary of expected impacts. This arises from the technology's versatility. Through genetic engineering, not only will totally new products become available, such as vaccines and pharmaceuticals, but alternative methods of production will be developed for many existing products. The impacts will clearly vary from product to product, process to process. An assessment of rDNA technology is quite different from an assessment of a technology such as off-shore oil drilling. In the latter, case, there is a single product, oil. One can examine the impacts of increasing oil supplies from the ocean—how the environment might be affected, how job availability will change, how the price of energy wil rise or fall, and so forth. In the former case, each application of rDNA technology will have its own spectrum of direct and indirect impacts. The impact of making a malaria vaccine available to the several hundred million at-risk people will have a totally different set of consequences than if a vaccine against hepatitis is developed. The populations affected by each vaccine differ; the economic impacts will differ, as will the social, political, and other consequences. This lack of generalizability applies to all products spanning a variety of industries already described. The fact is that entirely new avenues have been opened for the solution of society's problems.

Predictions are always dangerous exercises. Clearly, the prospects for biotechnology are enormous, and opportunities for innovation abound. Whether any one product or process falls into the realm of biotechnology may be debated. But biology has proved its value as a tool. It's almost a certainty that its impacts will be extraordinary. It is highly likely that from now on whenever a corporate manager or director of research needs to solve a product or process problem, he will consider a biological solution as one option. Biology has transformed society's way of thinking. And recombinant DNA was the catalyst.

Notes

1. U.S. Congress, Office of Technology Assessment, *Impacts of Applied Genetics* (Washington, D.C.: GPO, 1981).
2. U.S. Department of Agriculture, *Planned Genetic Resources: Conservation and Use* (Washington, D.C.: GPO/U.S. National Plant Genetic Resources Board, 1979); G. F. Sprague, D. E. Alexander, and J. W. Dudley, ''Plant Breeding and Genetic Engineering: A Perspective,'' *Bioscience* 30 (January 1980): 17–21.
3. D. Perlman, ''Fermentation Industries—Quo Vadis,'' *Chem Tech-Chemical Technology* 7 (1977): 434–43.
4. T. E. Wagner, P. C. Hoppe, J. D. Jollick, D.R. Scholl, R. L. Hodinka, ''Microinjection of a Rabbit Beta-Globin Gene into Zygotes and Its Subsequent Expression in Adult Mice and Their Offspring,'' *PNAS* 78 (1981): 6376–80.
5. Morris Levin, personal communication.
6. Ibid.
7. F. Lim and A. M. Sun, ''Microencapusulated Islets as Bioartificial Endocrine Pancreas,'' *Science* 210 (21 November 1980): 908–10.
8. G. Schmidt-Kastner, ''Biotransformation: New Directions for the Chemical Industry,'' *Robert S. First Conference on Biotechnology*, 11 May 1981.
9. Ronald Cape, personal communication.
10. Irving Johnson, personal communication.
11. J. K. Christman, M. Gerber, P. M. Price, C. Flordellis, J. Edelman, and G. Acs, ''Amplification of Expression of Hepatitis B Surface Antigen in 3T3 Cells Cotransfected with a Dominant-Acting Gene and Cloned Viral DNA,'' *PNAS* 79 (March 1982): 1815–19.
12. W. O'Neill, ''Implications of Molecular Genetics for Medicine,'' Office of Technology Assessment Contract Report No. 033-0980, June 1980.
13. Ibid.
14. J. A. Deutsch, ''Bombesin—Satiety or Malaise,'' *Nature* 285 (19 June 1980): 592.
15. W. O'Neill, ''Implications of Molecular Genetics for Medicine.''
16. R. S. Goodrich, ''Proalcool: Brazil's Large Scale Renewable Energy Program,'' *International Conference on Cybernetics and Society of IEEE, Session on Large Scale Renewable Energy Systems*, Atlanta, Ga. 26–28 October 1981.
17. Genex Corporation, ''Impact of Genetic Engineering Technology on Industry,'' Office of Technology Assessment Contract Report no. 033-1330, 15 April 1980.
18. Ibid.

18. Ibid.

19. R. F. Schleif and M. A. Favreau, "Hyperproduction of ara C Protein from *E. coli*," *Biochemistry* 21 (1982): 778.

20. William Amon, Cetus Corporation, personal communication.

21. "Socal Quits Venture with Cetus to Develop Fructose Process: Technical Snags Cited," *Wall Street Journal*, 2 June 1982.

22. Thomas Peyton, personal communication.

23. Genex Corporation, "Impact of Genetic Engineering Technology on Industry."

24. Council on Environmental Quality and Department of State, "The Global 2000 Report to the President" (1980).

25. J. Litchfield, "Microbial Protein Production," *BioScience* 30 (1980): 387–96.

26. I. Holzberg, "Engineered Foods," *Chemtech* 9 (February 1979): 110–13.

27. G. B. Carter, "Is Biotechnology Feeding the Russians?" *New Scientist* 89 (1981): 216–18.

28. Ibid.

29. Massachusetts Institute of Technology, "Impact of Applied Genetics on the Chemical Industry," Office of Technology Assessment Contract Report (1980).

30. D. Hall, M. Adams, P. Gisby, and K. Rao, "Plant Power Fuels Hydrogen Production," *New Scientist* 86 (10 April 1980): 72–75.

31. S. L. Gorbach, "Recombinant DNA, An Infectious Disease Perspective," *Journal of Infectious Diseases* 137 (May 1978): 615–23.

32. Ibid.

33. "Biotech Comes of Age," *Business Week*, 23 January 1984, p. 84.

34. David Baltimore, personal communication.

35. M. Cline, "Furor Over Genetic Engineering," *New Scientist* 88 (16 October 1980): 140.

The focus of the recombinant DNA controversy of the 1970s was clearly biohazards, and even that term has a somewhat restricted meaning as defined initially by the original concerns of scientists splicing genes into *E. coli* and wondering if a new source of human disease could be created. There are, of course, many other issues relevant to genetic manipulation—an expanded view of biohazards to include ecological disruptions from modified soil bacteria, for example; the ethics of human genetic intervention; the question of justice in the distribution of the fruits of biotechnology; the setting of priorities for the applications of the technology—for profit or human needs; the possibility of using the new techniques to create biological weapons. How these issues should be treated goes well beyond the matter of human safety in gene-splicing experiments.

Sheldon Krimsky, a social scientist at Tufts University, began his involvement in the policy aspects of recombinant DNA as a member of the Cambridge Experimentation Review Board, charged with examining the matter of gene splicing at Harvard and recommending local policy to Mayor Alfred Vellucci and the city fathers. Later he became a member of the RAC, when it was decided to expand the number of public members. Here he takes us out of the parochial context of the recombinant DNA controversy of the 1970s and asks us to examine the more profound implications of this technology as it matures in the 1980s.

12

Social Responsibility in an Age of Synthetic Biology: Beyond Biohazards

SHELDON KRIMSKY

The advent of gene splicing techniques was followed by ten years of caution in molecular genetics that focused mainly on the problems of potential biohazards arising from laboratory work. The response to a cornucopia of conjectured risks of gene splicing has not been trivial. New relationships have been created between science and its social institutions. These relationships are not only unique to the field of biology, but some are unprecedented for the entire enterprise of science. Among the changes that have taken place in the practice of DNA science are the following:

1. Biologists engaged in recombinant DNA research are accountable to other individuals or institutions for the safety practices in their laboratories involving the use of microorganisms.
2. A government agency established a special office and a panel consisting of biologists, ethicists, environmentalists, and public health professionals to oversee the risk assessment of rDNA research and issue mandatory guidelines to its grantees.
3. Local institutional biosafety committees were created with community representation to oversee rDNA work.

4. Laws were enacted by eight local governments and two states that regulate the use of rDNA technology for research and commercial institutions.

Some, if not all, of these institutional changes may only be temporary. During 1982 meetings, the NIH Recombinant DNA Advisory Committee (RAC) considered doing away with institutional biosafety committees (IBCs) and changing the mandatory NIH guidelines into a voluntary code of practice.[1] Although the National Institutes of Health has not completely disengaged itself from regulating the research in genetic engineering that it funds, responsibility for laboratory safety is gradually returning to the principal investigator where it had been prior to 1974.

Despite fears by some biologists, the involvement of nonscientists in the rDNA episode has not impeded scientific inquiry in any significant way.[23] But neither is there evidence that public participation on the RAC has contributed much to protecting the public health and welfare or to improving the decision-making process, although one would like to think that it had. Nevertheless, the process has been informative. New institutional mechanisms were developed to respond to a crisis in science over the safety of research. However we wish to interpret their effectiveness, the institutional arrangements reflect a heightened responsibility of science and government to society. A system of social guidance, steered by scientists, but open to public involvement, undertook the difficult task of trying to assess the laboratory hazards of the current use and future developments of a new technology.

The preponderance of attention given to laboratory safety has masked other important societal concerns regarding the commercial and military applications of rDNA technology. My purpose in this chapter is to draw attention to the potential impacts of gene-splicing technology that take us beyond the inadvertent creation of hazardous chimeric organisms.

In sketching out these issues, I shall not be seeking resolutions. I shall even suspend judgment on whether the concerns are ripe for a public response at this time. As with science, not all public policy dilemmas have matured sufficiently to warrant a response. The issues I shall raise about rDNA technology are more basic than a list of actual or hypothetical concerns about its social, economic, or environmental impacts because they question the context within which critical decisions are made.

What, if any, institutions are in place that can address social concerns associated with the new genetic technology as those problems arise? If such institutions or institutional responses are in place, are they appropriate to meet the demands of the problems? If there are no suitable institutions to deal with a generic class of problems, do we need any? What type(s) of institutions would be appropriate to satisfy public concerns? Is our society compatible with such institutions?

Beyond the issue of biohazards, I have identified five areas of concern most frequently expressed by the lay public, scientists, the media, and public interest groups.

1. What assurances are there that rDNA technology will not be used by the Department of Defense (DOD) to produce new biological weapons or improve conventional ones?
2. Now that an area of biomedical research has important social applications, why doesn't our government establish priorities for harnessing rDNA technology so that the greatest good for the greatest number can be sought? Is there any justice in allowing the free market to determine whether and to what extent gene splicing improves peoples' living conditions and to set the priorities for products introduced into the marketplace? It was, after all, social capital in the form of public funds that supported the basic research that eventually spawned rDNA technology.
3. Are we adequately protected from the inadvertent secondary impacts of new products and processes that commercial gene splicing can be expected to yield?
4. What safeguards are in place to protect society from the potential misuse of rDNA technology in human genetic engineering? Are there any ethical thresholds that should be considered when applying gene splicing or other forms of genetic engineering in the treatment of disease or in conjunction with other reproductive technologies? Should limits be set beyond which clinical research in human genetic engineering becomes impermissible?
5. Among the important applications of rDNA technology, some are particularly targeted to the Third World. Do we bear a responsibility to prevent the export of biotechnologies or their products that are either restricted, rejected, or simply not considered for use in our own country?

These issues are broadly framed and conspicuously unspecific. They form part of the critical rhetoric directed at the field of biotechnology. I shall elaborate on these questions in the process of examining the nature and adequacy of our institutional capacity to respond.

Recombinant DNA and Biological Weapons

The feasibility of creating biological weapons with rDNA technology was on the minds of the scientists who attended the Asilomar conference in 1975. Despite expressed concerns by some participants, the issue was kept off the agenda by the conference organizers for fear it would interfere with the primary goal of reaching a consensus on laboratory biohazards.[4] Nevertheless, one of the three working panels at Asilomar concluded its report on the assessment of risks with the following admonition:

> We believe that perhaps the greatest potential for biohazards involving genetic alteration of microorganisms relates to possible military applications. We believe strongly that construction of genetically altered microorganisms for any military purposes should be expressly prohibited by international treaty. . . .[5]

Just a few months prior to Asilomar, the U.S. Senate ratified the 1972 Convention on the Prohibition of the Development, Production, and Stockpiling of Bacteriological and Toxin Weapons. The Convention's articles were put into force in this country by March 1975. Nearly a hundred countries had already pledged "never in any circumstances to develop, produce, stockpile or otherwise acquire or retain (1) microbial or other biological agents . . . whatever their origin or method of production, of types and in quantities that have no justification for prophylactic, protective or other peaceful purposes; (2) weapons, equipment or other means of delivery designed to use such agents or toxins for hostile purposes or in armed conflict."[6]

The Convention requires a review after five years to examine the relevance of new scientific and technological developments. A review committee report issued in March 1980 concluded that biological materials constructed by rDNA techniques were unambiguously covered in the Convention's language. The committee also assessed the potential use of rDNA for creating biological weapons. The prospect of developing fundamentally new agents or toxins with rDNA technology was viewed as a problem of "insurmountable complexity." The committee saw little incentive for such efforts since "naturally occurring, disease-producing microorganisms and toxins already span an exceedingly broad range, from some which are extraordinarily deadly to others usually producing only temporary illness." However, the committee considered more probable the use of rDNA techniques to improve the effectiveness of existing biological warfare agents.[7] (For an extended discussion of recombinant DNA research and biological warfare, see Chapter 10.)

What assurance does the public have that present or future administrations will adhere to the articles of the Convention? What institutions are currently available to provide the public with information on DOD biological research programs? Is there a clear distinction between offensive and defensive biological weapons? Does it even make sense to speak about defensive biological weapons? What would they be and is their production restricted by the Convention?

As an example, might the army want to clone toxigenic genes into wild type *E. coli* to assess the potential use of gene splicing by a hostile state or terrorist group? According to a 1980 report, DOD has expressed an interest in assessing whether genetic engineering can be used to make biological weapons:

> New threats may be opened up by various technological and scientific advances. As examples, recombinant DNA technology could make it possible for

a potential enemy to implant virulence factors or toxin-produci
mation into common, easily transmitted bacteria such as *E. c*
context, the objective of this work is to provide an essential b
information to counteract these possibilities and to provide
standing of the disease mechanisms of bacterial and rickettsia
pose a potential BW threat, with or without genetic manipulation.[8]

To improve its understanding of these possibilities the U.S. Army has begun some rDNA work. Its Medical Research Institute of Infectious Diseases received permission from the RAC to clone *Pseudomonas* exotoxin in *E. coli*.[9] The Army Medical Research and Development Command advertised in *Science* for proposals on the introduction by rDNA methods of the human nervous system gene acetylcholinesterase into a bacterium.[10] The expressed purpose of the research is to develop an effective antidote for nerve agents manufactured by the USSR that are extremely potent cholinesterase inhibitors. The Army's interest in cloning the enzyme is to obtain a sizable quantity of high-purity material so that its physical and biochemical properties can be studied.

What windows of accountability exist between the military and the public on the uses of genetic engineering? How can public skepticism be turned into public confidence? Presently, there are three institutional responses that serve to build public confidence: the 1972 Convention previously mentioned, a federal law requiring the DOD to describe its obligated funds in chemical and biological research, and the NIH guidelines. The second institutional response gives Congress and the public more direct access to the chemical and biological research carried on by the military. According to Public Laws 91-121 passed in 1969 and amended in 1975 (P.L. 93-608), DOD is required to submit an annual report to Congress that explains obligated funds in its chemical warfare and biological research programs.

The third instrument of social accountability is the NIH guidelines. A memorandum from the Undersecretary of Defense dated 1 April 1981 states that "all DNA activities funded by DOD, whether in-house or by contract or grant, will be conducted in full compliance" with the NIH guidelines.[11] The ruling specifies the use of institutional biosafety committees and requires that a complete file of each research project be maintained for public scrutiny at the U.S. Army Medical Research Institute of Infectious Diseases.

It appears then, that if the army wishes to clone toxigenic genes into wild type *E. coli*, current DOD policy requires that these experiments must first receive approval from the RAC and subsequently be registered. There is no reason to believe that DOD will not adhere to the letter of the Convention. At issue here are the bridges of confidences that must exist between the public and the military.

If the NIH guidelines become significantly relaxed or are turned into a voluntary code of practice, neither the RAC nor the local institutional biosafety committees will continue to provide the public with an important window of accountability into the military uses of rDNA technology.[12]

Priorities for Harnessing rDNA Technology

Shall the government play a role in directing the application of rDNA technology, either by setting national priorities, establishing incentives to promote specific commercial developments, or by directly funding the development of high-priority products and processes? The alternative is to let the free market determine the priorities. Three arguments have been advanced to support a strong governmental role in exploiting the social uses of genetic engineering.

Argument 1. Since public monies were the principal source of funding through which rDNA methods were developed, the public sector should play a major role in directing its use. A corollary to this position states that the public is entitled to a return on its investment and should control the patents on products and processes that grow out of federally funded research.

Argument 2. If social priorities are not set for the use of rDNA technology then the public will miss out on important applications that private markets will not find profitable to pursue. A case in point are "orphan drugs," which illustrate the need for governmental involvement in the development of pharmaceuticals. The drugs are so named because they cannot find a "parent" company that will invest in their manufacture. The markets for the drugs are perceived as too limited to provide a satisfactory return on investment. Few question, however, the responsibility of society to make available for clinical use nonprofitable drugs if they are effective in aiding even a small number of patients.

Argument 3. When the benefits of rDNA technology are realized, such as in agriculture, it is the responsibility of government to guide the benefits so that they are at least shared equitably and at most shared in a manner that narrows distributional gaps. In the case of agricultural impacts, a guidance system can ensure that small farmers are not disadvantaged from new strains of genetically engineered seed stocks, that the consumer gets a better quality product at a more reasonable price, and that environmental health is not traded off for higher rates of return.

For the purpose of this discussion I shall suspend judgment on the cogency of the arguments. They do, however, form an essential part of the background criticism that has been raised against the fledging gene-splicing industry, in which hundreds of biotechnology firms have surfaced in a highly competitive marketplace with their own sets of agendas and perceptions of social needs. Returning to my initial query, Are there institutions that can establish priorities for harnessing recombinant DNA technology? Can some institutional response insure that the private and public investments in the field of biotechnology get transformed into uses that are responsive to distributional inequities?

Currently, no single institution has the authority to set and implant priorities for the application of gene-splicing methods to industrial, agricultural, or clinical areas. Moreover, there is little, if any, precedent in this country to guide the development of a technology of such broad scope. While we have governmental institutions for setting research agendas, assessing and controlling technological impacts, and overseeing targeted programs in applied technology, private markets are fundamentally responsible for what gets produced, in what order, and toward what end.

In theory at least, the Office of Technology Assessment (OTA) possesses an excellent vantage point from which to establish a set of priorities for developing rDNA technology. But on the basis of its evolving role over the past several years, which excludes advocacy of particular policies and actions, it is highly unlikely that OTA would be the body to establish a hierarchy of needs from which to develop a strategy for extracting social benefits from the technology. Different agencies of government such as the Department of Energy (DOE), the Food and Drug Administration (FDA), and the Department of Agriculture (USDA) will set their own agendas. However, the public has little understanding and input in how these agencies establish their individual priorities.

For example, the Plum Island Animal Disease Center (USDA) and the bioengineering firm Genetech, Inc., have entered into a cooperative agreement to produce a vaccine for foot-and-mouth disease. For many countries outside the United States, the highly infectious foot-and-mouth virus is responsible for substantial losses in beef stock and milk production. Fortunately, North American agribusiness has been spared the disease for many decades because of strict beef import controls. It is argued that American consumers could benefit from a world-wide eradication of the disease. For Genetech, the carrot in this public-private partnership is the right to foreign markets for vaccine sales (see Appendix), a sizable benefit for a small firm that entered the vaccine research program in its final stages.

Broader public input for setting agency priorities could come from Congress through its appropriate subcommittees on science and technology. In 1981, OTA offered several options for Congress to consider with regard to promoting advances in biotechnology. The options included establishing a funding agency in biotechnology, creating federally financed research centers in universities, providing tax incentives to expand the capital supply to small high-risk firms, improving conditions under which U.S. companies collaborate with academic scientists, and mandating support for specific research programs.[13] Each of these recommendations has been used in the past to stimulate or set a direction for the development of innovative technologies. But the approach taken by OTA does not address the question of setting overall priorites for the utilization of rDNA technology on the basis of social needs. Precedents in this country run against this type of endeavor, which might be termed national technology planning.

The government has taken an active role in shaping the direction and quality of research, both targeted and basic, through funding mechanisms.[14] It is estimated that about two thirds of all the basic research carried on in the U.S. is federally supported. But the public sector role has been minimal to almost nonexistent in directing the application of technologies. It is widely assumed that social needs will be more effectively revealed through market forces. Notwithstanding the sudden growth of economic fundamentalism, there are many areas where the assumption fails miserably. A former member of the White House Office of Science and Technology Policy and an astute observer of genetic technology offered this prognosis:

> There is certain to be a shake-out in biotechnology over the next five years or so, and the determinants of success are likely to be related more to business strategies, shrewd management, and high quality control, assuming a strong base of laboratory talent. Chemical and agricultural products will be marketed in short order, if they are economically competitive, but hormones, drugs, and vaccines must undergo complex and expensive clinical trials. There can be little doubt that the Wall Street criteria will be applied: earnings, growth, profitability.[15]

Since gene splicing is expected to introduce innovations in several commercial fields and if a hierarchy of needs is conceded to be desirable, it seems reasonable that it be achieved within specific areas, such as vaccine production, chemical processing, biomass conversion, and agricultural products. Federal agencies such as FDA, DOE, and USDA could set priorities in biotechnology with appropriate inputs from the public. It has also been suggested that "we might set as a national goal the conversion of an economy based on fossil fuel to an economy based on microbial fermentation."[16] A national effort, on the scale of the space program, that promotes the development of inexpensive, renewable energy sources from biotechnology would improve significantly the public's confidence in science and technology.

Who decides what gets manufactured first, bovine growth hormone to fatten up cattle or human growth hormone as a replacement therapy for a genetic defect? Should a vaccine for malaria get precedence over one for herpes virus? Other factors besides profit and social demand will invariably enter into such determinations, such as how far advanced the state of knowledge is toward a solution of a particular genetic engineering problem. Considerations also include what sources of private and public capital are available for specific product development. Products with low market potential are likely to be left behind. There are no established institutions or advocacy networks through which the public sector can make its voice heard on priorities in technological innovation. Yet the public has been promised so much from biotechnology in such a short time that federal responsibility and initiative in supporting a development program deserve careful consideration. The first step is to recognize the legitimacy of the citizenry in steering a technology. The next step is to develop avenues of participation and social guidance mechanisms.

Secondary Impacts

The products, processes, and industries that eventually emerge as a consequence of the commercial applications of molecular biology will undoubtedly exhibit unintended secondary impacts. For example, if a vast array of new and old pharmaceutical products are manufactured at low cost (including antibiotics and insulin), what effect, if any, will that have on drug overuse?[17] If agricultural plants are engineered to be resistant to herbicides, will that stimulate a greater use and dependency of chemicals in agriculture? If microbes are developed to degrade herbicides containing dioxins, will that justify the removal of restrictions on this class of potent chemicals if they are used in conjunction with their biological antidote?

It is easy to raise hypothetical cases. I do not pretend to have any skills as a technological forecaster. I use these not-so-implausible cases to reintroduce the theme of my inquiry, Are there social guidance systems through which society can anticipate, or at least keep track of, the secondary consequences of major technological innovations?

In the late 1960s, the term technology assessment became a part of our policy vocabulary. The National Environmental Policy Act established a requirement for environmental impact assessments for many government supported projects. Even the NIH guidelines for recombinant DNA research were issued with an environmental impact statement.[18] Some laws and institutions are already in place to respond to the potential impacts of biotechnology. None, however, are equipped to provide continuous monitoring and assessments of the full range of expected commercial uses of gene splicing and hybridoma technology over a period of years.

In principle, the Office of Technology Assessment is well suited to carry out this function because it has been able to assemble highly trained interdisciplinary teams. But as an agency of Congress, OTA will undertake studies only when there is sufficient congressional support. In its 1981 study on the impacts of applied genetics (see note 13), OTA chose to place considerable weight on the positive social aspects of genetic engineering, including increased manpower needs, new products, and improved yields in agriculture. But only a feeble effort was made in this study to evaluate potential adverse social or ecological consequences of industrial microbiology. No research program or framework for long-term assessment is offered.

A second study by OTA completed in 1984 evaluated the competitive position of the United States in the global biotechnology field.[19] In the scope of project activities there is a striking absence of any reference to the assessment of secondary impacts. For OTA, technology assessment in the field of biotechnology has come to mean promotion, pure and simple.

Individual agencies also bear responsibilities under legislative mandate to evaluate products of rDNA technology. But evaluations of this nature are spotty and restricted in scope. The FDA is primarily concerned about the efficacy, purity, and the side effects of a new drug, but cannot rule on its broader social manifestations. The Environmental Protection Agency (EPA) does have authority over the release of hazardous materials into the environment. Currently, no regulations exist that restrict the release of biological agents into the environment with the sole exception of biological pesticides, which must be evaluated for safety and efficacy according to the Federal Insecticide, Fungicide, and Rodenticide Act (FIFRA) before they are registered. However, in 1984 EPA published a notice of intent to promulgate regulations covering the release of genetically modified organisms. The agency cites the 1976 Toxic Substances Control Act as providing the legislative authority for regulating microorganisms.

Although EPA initially planned to issue regulations for infectious wastes under the Resource Conservation Recovery Act, the agency has changed its approach under the Reagan administration. Its current effort is directed at publishing a set of voluntary guidelines in the form of an infectious waste management plan.

It is clear that biotechnology is today where the petrochemical and nuclear technologies were forty years ago. The experience we gained in these fields should not be neglected. In the absence of adequate guidance systems the social consequences of technology come without advance warning and in a form in which effects are all too often irreversible.

Human Genetic Engineering

The discovery of gene splicing as a tool of scientific inquiry received considerable media attention because of the initial concern about producing hazardous organisms. But even as these issues were debated, sectors of society began to draw attention to human genetic engineering. States and local communities that took up the regulatory issues were confounded by the ethical ones, if they considered them at all. The exception is Waltham, Massachusetts, which passed a law that forbids the use of humans in rDNA experiments. The human-experiment provision, tagged on to its ordinance regulating rDNA activities, did not result from a broad community debate, but was prompted by a single member of the city council.

It is not difficult to conjure up insidious forms of human genetic manipulation. Nor is it problematic to imagine humane treatments. But there is a vast middle ground on the use of genetic engineering for which a consensus does not exist among scientists, ethicists, members of the religious community, or the general public. Some view alteration of germ-line cells as morally reprehensible. Others argue that we have just as much of an obligation to eliminate from the gene pool the determinants of Tay Sachs and sickle-cell anemia as we have to eliminate smallpox from the planet. Germ-line cell surgery may be the only way to achieve such a goal.

In 1977, at the National Academy of Sciences' Academy Forum devoted to genetic engineering, a scientist tried to put into perspective both the lofty claims and the exaggerated fears being expressed about rDNA technology. Questioning the promise of genetic surgery he said, "How about a thalassemic? Are we going to drain his marrow out, then culture his cells, get DNA in [the cells], and put [them] back in [the person]? Quite frankly I would rather be a thalassemic than have happen to me."[20]

However incredulous the implantation of genetically engineered cells may have appeared at the time, just three years after the Academy Forum an investigator at the University of California at Los Angeles performed a remarkably similar procedure in what has been deemed the first human genetic engineering experiments. The two subjects, who voluntarily consented to the experiments, were from Israel and Italy. Similar protocols were not approved at UCLA at that institution's human experimentation committee. The subjects in question were suffering from a life-threatening blood disease, beta thalassemia major, in which the bone marrow cells are unable to produce normal hemoglobin because of defective genes.

The UCLA investigator removed bone marrow cells from the patients and exposed the cells to normal genes for making hemoglobin. The human genes were cloned by rDNA techniques. The gentically engineered bone marrow cells were reinjected into the patient in the hope that they would prosper and produce blood cells with normal hemoglobin. The individuals treated by this procedure were experiencing the final stages of the disease and given the prognosis of a limited life expectancy. UCLA's human subjects's committee, after holding the application for fourteen months, was unwilling to approve the experimental procedure on the grounds that there was insufficient evidence of its success in animal systems.

What can be said about the current institutions available to handle such questions? Committees for the protection of human subjects, whose operating procedures are defined in federal guidelines, issue independent judgments at each institution. But human genetic engineering represents a quantum leap in the use of humans as experimental subjects. There are more issues at stake than the protection of the rights of privacy and well-being of individual subjects.

In cases where the genetic engineering is performed on the fertilized egg in vitro (combining genetic engineering with in vitro fertilization), review by human subject committees is not required. It is conceivable that the RAC could address the issue of human genetic manipulation, but its current charter and membership is designed to keep RAC's attention exclusively to biohazards and away from ethical problems.

There is another institutional process for reviewing human genetic engineering experiments in the President's Commission for the Study of Ethical Problems in Medicine and Biomedical and Behavioral Research. Since the Commission can only issue recommendations, it cannot provide the oversight to local human subjects committees unless Congress acts or some initiatives are taken by the Department of Health and Human Services, which oversees the human subjects regulations. Without some general guidelines establishing ethical norms for human genetic manipulation experiments, the or developing such policies will be relegated to the individual mmittees. And while the possibilities for human genetic ther- icing in conjunction with in vitro fertilization, and mamma- ow, free-floating public anxieties are in search of an institu-

Biotechnology and the Third World

The final area of social responsibility in my brief survey pertains to the transfer of biotechnology to the Third World. The responsibility is two-sided. On one hand, the industrialized world must find ways to share the positive fruits of genetic engineering with developing nations without destroying their unique cultural forms and neglecting their capacities for self-determination, or the needs of their economic systems. On the other hand, we bear an obligation to prevent the export and development of products and processes that we determine to be unfit for ourselves or that would be unsuitable for the cultural patterns and technological development of the country in question.

Very soon after plasmid-mediated transfer of chimeric genes was discovered at Stanford, its commercial applications were being seriously investigated. One scientist, who was employed by a major U.S. corporation, saw in genetic engineering a solution to the problem of world hunger. This scientist was planning to construct a plasmid with genes from *Pseudomonas* that code for enzymes with cellulose-degrading properties. He considered cloning the plasmid with the celluose-degrading genes in *E. coli*. His logic for using this organism to alleviate world hunger was as follows. Suppose that vast populations in underdeveloped food-scarce countries could have their intestinal flora transformed or replaced with the new cellulose-degrading *E. coli*. With their new intestinal flora, these individuals could presumably obtain caloric value from vegetation that is plentiful and inexpensive, but under present circumstances nutritionally useless to them. After being advised by a scientific colleague that cellulose-degrading *E. coli* in the gut could eliminate any roughage in the digestive tract and thus increase the rates of certain disease correlated with low-fiber diets (obesity and bowel cancer were cited), the investigator gave up his plan.[21]

I present this story not to emphasize the potential hazards of such a scheme nor to question the motives of responsible scientists, but to remind us that the idea was being considered for use exclusively in the poorer nations of the world. Currently, transnational corporations market and export products to developing countries that are either prohibited or severely restricted for domestic use. Publications and television documentaries have illustrated these problems for the export of pesticides and pharmaceuticals.[22] Currently, there are no federal agencies with authority to prohibit the export of domestically banned products even when they are manufactured in this country. Under the Federal Insecticide, Fungicide, and Rodenticide Act and the Toxic Substances Control Act (TSCA), there is a statutory obligation to inform foreign governments of regulatory actions taken against specific products. Under TSCA, companies are required to make toxicity information on chemical products available to foreign import countries. Notwithstanding the legality of the transactions, and those cases of domestically banned exports whose benefits to the import country clearly outweigh their adverse effects, there remain many areas where the ethical practices of the exporting firms are questionable.

Just prior to leaving office, President Carter signed an executive order (15 January 1981) that placed export controls on extremely hazardous substances deemed a substantial threat to human health, safety, or the environment. Once classified, these products would require an export license. Carter's order took two years to develop. It was in effect for only a month when it was revoked by President Reagan.[23]

The first commercial products of rDNA technology are already reaching consumers. It is reasonable to expect the problem of questionable exports to be compounded as biotechnology's pioneers seek world markets. For example, since pesticides contaminated with dioxins are already being sold to some developing countries, there is also a market for the biological antidote—organisms genetically engineered to degrade this class of pesticides. The media, reporting the development of a new genetically engineered bacterium that degrades the herbicide 2,4,5-T, quoted its creator (who happened to have received the first patent for an oil-eating bacterium) as saying, "If you use 2,4,5-T to kill weeds one year and then apply these microorganisms to clean [up the 2,4,5-T] . . . then their numbers will die out drastically when their (sic) is no more of the chemical to eat."[24] Under such ideal but unrealistic conditions, the microbes will not mutate, establish themselves in new niches, or adversely affect the microflora of the land areas on which they are sprayed. Less dubious assumptions have been responsible for creating havoc in sensitive ecological systems.

Our current experience with exports of hazardous materials and the transfer of certain inappropriate technologies to industrially underdeveloped agrarian societies leads me to the conclusion that unless more responsible institutions are created, similar mistakes will be made as the fruits of rDNA technology are realized. The widely used justification that importing countries are free and willing to buy products of dubious value fails the test of moral reciprocity. The Carter executive order was a positive step toward a global responsibility for our exported products. With that order rescinded, an additional burden is placed on scientists and members of the public health community who are familiar with the deleterious effects of chemical exports to inform the recipient countries and the World Health Organization before rather than after severe injury or environmental damage occurred.

The medical and commercial applications of gene splicing raise some old problems in new clothing. The developments in a field bursting with innovative ideas and unlimited potential will put to the test the social guidance systems we presently have. But especially, they will test the moral and scientific wisdom of technologically advanced countries on their capacity to counteract the adverse possibilities of genetic technology before they are realized and become part of the social and economic fabric of society.

Scientific knowledge, said Bacon, is power, and power comes in two denominations, liabilites and assets. It is unthinkable that nature will release its genetic genie in only a single denomination. That type of technological accounting results in moral bankruptcy. The key to sound technological bookkeeping is in the development of guidance systems whose sole function is to troubleshoot biotechnology's social impacts. Efforts toward this goal have been notable in the areas of laboratory safety and military applications. Initiatives in other areas remain less than satisfactory.

Notes

1. Marjorie Sun, "DNA Rules Kept to Head Off New Laws," *Science* 215 (26 February 1982): 1079–80.
2. Maxine Singer, "Spectacular Science and Ponderous Process," editorial, *Science* 203 (5 January 1979): 9.
3. Jane Setlow, "How the NIH Recombinant DNA Molecule Committee Works in 1979," *Recombinant DNA and Genetic Experimentation*, in J. Morgan and W. J. Whelan, eds. (Oxford: Pergamon Press, 1979), 161–63.
4. See, for example, Sheldon Krimsky, *Genetic Alchemy: The Social History of the Recombinant DNA Controversy* (Cambridge, Mass.: MIT Press, 1982), p. 106.
5. Plasmid Working Group, *Proposed Guidelines and Potential Biohazards Associated with Experiments Involving Genetically Altered Microorganisms*, 24 February 1975, Institute Archives and Special Collections, MIT, p. 19.
6. *Bacteriological (Biological) and Toxin Weapons*, Convention between the United States and other governments, 10 April 1972, at Washington, London, and Moscow (Washington: D.C.: GPO, 1975), p. 5. Under executive order, offensive biological warfare research was terminated in the United States in November 1969.
7. *Report of the Preparatory Committee for the Review Conference of the Parties to the Convention on the Prohibition of the Development, Production, and Stockpiling of Bacteriological (Biological) and Toxin Weapons and on Their Destruction*, 1980, p. 7. UN Document BWC/Conf. 1/10, 21 March 1980.
8. Department of Defense, *Annual Report on Chemical Warfare and Biological Research Programs* (1 October 1979–30 September 1980), 15 December 1980, Sec. 2, p. 4.
9. Nicholas Wade, "BW and Recombinant DNA," *Science* 208 (18 April 1980): 271.
10. U.S. Army Medical Research and Development Command, "Requests for Proposals," *Science* 209 (12 September 1980): 1282.
11. James P. Wade, Jr., Acting Under Secretary of Defense, memorandum to assist-

ant secretaries Army, Navy, and Air Force, 1 April 1981.

12. The Office of Recombinant DNA Activities maintains a list of institutions that have registered IBCs with NIH. The list includes governmental agencies and private firms under the voluntary compliance program. As of February 1984, the registry included the Walter Reed Medical Research Institute of Infectious Disease, Washington, D.C., and the U.S. Army's Chemical and Biological Division, Frederick, Maryland. The Naval Bioscience Laboratory in Oakland, California, was noticeably missing from the list.
13. Office of Technology Assessment, *Impacts of Applied Genetics: Microorganisms, Plants, and Animals* (Washington, D.C.: GPO, April 1981), 10–12.
14. See, for example, T. Cooper and J. Fullarton, "The Place of Biomedical Science in National Health Policy," in *Biomedical Scientists and Public Policy*, H. H. Fudenberg and V. L. Melnick, eds. (New York: Plenum Press, 1978), 132–52.
15. Gilbert S. Omenn, "Government as a Broker between Private and Public Institutions in the development of Recombinant DNA Applications," *Proceedings, Genetic Engineering International Conference, Vol. 1* (Seattle, Wash.: Battelle Memorial Institute, 1981), p. 44.
16. Herman W. Lewis, "Role of Government in Promoting Innovation and Technology Transfer," *Proceedings, Genetic Engineering International Conference, Vol. 1* (Seattle, Wash.: Battelle Memorial Institute, 1981), p. 46.
17. Ruth Hubbard, of the Harvard Biology Department, drew the following conclusion in her assessment of the use of rDNA technology for the production of insulin: "Given the history of drug therapy in relation to other diseases, we know that if we produce more insulin, more insulin will be used, whether diabetics need it or not." National Academy of Sciences, *Research with Recombinant DNA* (Washington, D.C.: NAS, 1977), pp. 165–69.
18. United States Department of Health, Education and Welfare-NIH, *Environmental Impact Statement on NIH Guidelines for Research Involving Recombinant DNA Molecules*, October 1977, in two parts.
19. Office of Technology Assessment, *Commercial Biotechnology: An International Assessment* (Washington, D.C.: GPO, January, 1984).
20. National Academy of Sciences, *Research with Recombinant DNA* (Washington, D.C.: NAS, 1977), p. 170.
21. Reconstructed arguments from the interchange of this correspondence are presented in Krimsky, *Genetic Alchemy*, pp. 117–119.
22. David Weir and Mark Shapiro, "The Circle of Poison," *The Nation* 231 (15 November 1980): 514; Francine Schulberg, "United States Export of Products Banned for Domestic Use," *Harvard International Law Review* 20, no. 2 (Spring 1979); Public Broadcasting Service television broadcast, "Pesticides and Pills: For Export Only," aired 5 and 7 October 1981.
23. Rashid Shaikh and Michael R. Reich, "Haphazard Policy on Hazardous Exports," *Lancet*, 3 October 1981, pp. 740–42.
24. Philip J. Hilts, "New Chemical-Eating Germ May Gobble Up Toxic Waste Dumps," *Washington Post*, November 30, 1981.

Epilogue: Looking Ahead

BURKE ZIMMERMAN

This is a volume of observations and opinions. The actual facts as they happened can only be portrayed through the eyes of the several observers. Like the often-cited experiments on the eyewitness reports, we see that here, too, different individuals describe the same occurrences in very different terms. When it comes to interpretation and analyses of *why* events unfolded as they did, the variation is vast. But, then, the divergence of views and the explanation of events, even among a group of scholars whose profession it is to describe the Truth, has never been known to be less.

Once more, we are shown that the writing of history is a profoundly human endeavor. "Facts" are first perceived by humans who already hold varying opinions and biases, which modulate and change those events as they are written into their own memories. As they are retrieved and analyzed from time to time, they are further refined to fit into the worldview of the observer. Finally, the recollections and the attendant theories, interpretations, and conclusions are written down for posterity. Fortunately, there are usually a number of observers and historians for any notable event, or the students of the future would be in sorry shape. Of course, that still does not permit the selection of an account that is error free, or allows us to select the linear combination of all of them that would provide the best fit to reality. But as long as it is humans who write and interpret history, this is what we are stuck with.

In spite of this inexorable entanglement with human frailty, this volume nevertheless provides a collective, whole view of one of the most anguished and bitter controversies in modern science. When these chapters are taken together, the reader can obtain a flavor for the intensity and basic irrationality of the struggle and perhaps understand why everyone was indeed "rushing to tell the King." In this synthesis, can we now step outside of our immediate lives and the pressures and passions that drive them, and extract and define the nature of its legacy? For surely, all of that commotion could not have passed without somehow leaving its mark.

One thing it has done is perhaps not in our best interests. Many scientists who were interested in an initial dialogue on the matter of risk and the appropriate precautions to be taken with respect to research using recombinant DNA techniques see now that they had only hell to pay. Instead of commendations for their prudence and restraint, there were charges of a conspiracy of scientists to put one over on the public. Who would want to go through that again? But does this mean that when new discoveries suggest possible hazardous consequences, the discoverers will simply keep quiet and not mention that there could be new matters that merit public attention?

And do the tremendous advances in molecular and cell biology hold new causes for concern? Or does our considerable experience with the splicing of all manner of genes now tell us that it is as safe as any other endeavor? There is probably truth in both views. The ways in which novel genetic combinations are constantly being invented are always accompanied by questions about undesirable corollaries. As in the early cases of adding a small amount of DNA to *E. coli,* the probability of accidental problems is probably very small. But no one can say that it is zero. Now we may be faced with the deliberate construction of novel substances or modified organisms such as immunotoxins (a protein toxin linked to a monoclonal antibody, usually for the purpose of designing a specific therapeutic agent for cancer, or perhaps a biological pesticide) or strains of bacteria designed to modify agricultural soils. Like the pesticide panacea of the 1940s and 1950s, they may do more than they were designed to do.

The power of the knowledge of the details of genetic structure and function is being felt in other ways and they may have implications that far exceed the effects of genetic accidents. The problem is the manner in which the knowledge is used—for whom, by whom, and for what—as Sheldon Krimsky notes in Chapter 12.

Whether one is concerned about the manipulation of the human genome for nontherapeutic purposes, or the matter of equity in receiving the benefits of the prospective miracle cures of biotechnology, or the question of priorities—profit versus meeting real human needs—the conclusion is the same: The power of the technology is now so great that many believe its directions cannot be left to chance, or to free enterprise, or to military dictators looking for a more diabolical weapon. What is needed is the development of an ethic to govern technology, or at least to establish a process by which ethical considerations are taken into account. Here we are referring to biotechnology in its most general sense, but it is difficult to speak of a technology ethic without including all technology.

The questions begging for answers from such ethical analyses are many. How can potential military applications be controlled? Should the limited resources for exploiting the new biology be directed toward the solution of pressing human needs—hunger, tropical diseases, environmental destruction and pollution—or allowed to follow the dynamics of commercialism? While the private biotechnology industry is trying to find the bonanza of a cure for cancer (which afflicts about one tenth as many people as are afflicted with malaria), it may also apply biotechnology to the lucrative markets for novel noncaloric sweetners, cosmetics, food additives, and a longer lasting chewing gum. This is perhaps a reflection of the same commercial ethic that has found that the most profitable application of modern computer technology is video games. But there are many who are disturbed by the prospect of squandering on the frivolities of the affluent Western world a technology that surely can better the human condition. Or can we perhaps have our synthetic-peptide sweetened cake, with its cloned enzyme texturized dough, and eat it too?

There may yet be another bioscare. Chicken Little may once again scream of impending disaster and gather her allies to seek salvation from impending doom. There is even the chance that someday she may be right. But gene splicing and biotechnology are not novelties—they are realities with enormous potential. The possible directions in which the technology can be developed are numerous. The choice of those available will have a great effect upon our future. If the concept of preparing and planning for our future is at all a valid one, now is the time to begin.

Some steps are already being taken. Before its legal existence expired early in 1983, the President's Commission for the Study of Ethical Problems in Medicine and Biomedical and Behavioral Research published a report of its study on the ethics of human genetic intervention, called *Splicing Life*. The commission attempted to look well ahead of the capabilities of the current technology, to the day when the modification of the human genome becomes a reality. While questioning the desirability of human genetic modification for other than therapeutic purposes, the commission recommended that an oversight body be established in the United States, perhaps a permanent one, to keep track of the uses of this powerful technology and recommend policies to the government. Congressman Albert Gore, then chairman of the Subcommittee on Oversight and Investigations, of the Committee on Science and Technology, held a hearing on the subject and promised to take the initiative in proposing legislation for the establishment of such a body, although Congress has yet to take any action to form such a body. The matter is also being addressed by private scholars and organizations who believe the potential consequences of the technology are too important to be left to chance or the dynamics that have governed the development of technology in the past.

In much of the world, especially outside of the economically developed Western countries and Japan, there is concern that the problems of the developing countries will be ignored by a technology dominated by commercial incentives. There is, for example, little economic incentive to develop vaccines against schistosomiasis or malaria. Genentech dropped its sponsorship of a project with New York University to develop and produce a malarial vaccine, a project that had already received funding from the World Health Organization. The reason given was that Genentech felt that WHO's involvement made it impossible for them to have exclusive rights to sell the vaccine once developed. This is just one of many such ironies.

There is also little economic incentive to improve the crop varieties that feed most of the world. The American firms are concentrating on improvements in the largest crops in the United States—corn, soybeans, and cotton. Much of the world is also faced with poor soils or arid lands containing salt concentrations too excessive for productive farming. Unless these countries are willing to spend precious financial resources to allow commercial organizations to undertake the solution of such problems at a profit, they will receive no attention from these companies.

A number of developing countries, such as Pakistan, Thailand, and Egypt have begun to establish their own research and development institutes to address these problems. While they are generally far behind the research institutions in the developed countries, both in the quality of their facilities and in the level of skill of their scientists, some are notable for their successes. The International Rice Research Institute in Los Banos, the Philippines, has been in existence for over twenty years. Using conventional techniques in plant genetics, it has developed greatly improved strains of rice, which now make up a significant fraction of the rice grown throughout the world. Such institutions may provide the first successful applications of plant genetic engineering to the agricultural problems of the developing nations.

Since 1981, the United Nations Industrial Development Organization (UNIDO) has been working to establish the International Centre for Genetic Engineering and Biotechnology as a world-class scientific research, development, and training institution concentrating on the solution of the problems of developing countries. The initial efforts have been somewhat hampered by the usual politics, common to the proceedings of all international organizations, and most of all, the lack of sufficient resources.

The United States under the Reagan administration has rejected involvement in the establishment of the International Centre. American delegates were forbidden to attend the organizational meeting in Belgrade in December 1982. Apparently there was the worry that if the United States were to participate, it might give away some of its technology and weaken its world commercial competitive position or perhaps even its military position. There was an accompanying reluctance to commit even a few million dollars toward the establishment of such an effort, an amount miniscule in comparison with the level of foreign military aid. The current financial commitments of the United States to other countries are the response to a crisis-oriented foreign policy that sees ideological conspiracies rather than the vulnerability to corruption and exploitation that results from poverty and squalid living conditions. The Agency for International Development (AID), functioning somewhat independently, has been able to commit several million dollars toward the funding of biotechnology projects in developing countries.

Finally, what of the use of biotechnology for military purposes? The question has already been raised in the United States with regard to some of the novel technologies coming out of the new biology, such as immunotoxins, as Raymond Zilinskas has noted in Chapter 10.

These, then are some of the problems of the future. Some are monumental, both in scope and in the way they will influence our futures. If there is a lesson to be learned from the fussing and fighting of the 1970s over just how we would deal with a hypothetical hazard, it is that now is the time to behave like adults. We cannot shun difficult questions because they may generate controversy and possibly rancor. Before us now are a number of difficult questions arising from the brilliant advances in biological sciences, and we must not avoid them. They are tangible, real, well-defined questions, unlike the subject of potential hazards. We all have a stake in their answers.

We must define an ethic and priorities for an important technology. This will be the sequel to the "gene-splicing wars" of the 1970s; it will have many facets and a continuum of subissues. The new debates have begun in forums around the world. Former Congressman Gore received letters both supporting and warning against the establishment of a new oversight body on human genetic engineering. Already the fears of arbitrary congressional action are being expressed again. The National Research Council and the International Student Pugwash Conference have both addressed the matter of priorities for biotechnology as a world-wide problem. The problems of making the best use of a powerful technology to solve the most pressing world problems continue to be discussed in several branches of the United Nations.

We have thus struggled—or muddled—through the recombinant DNA controversy and have gained at least some measure of enlightenment. Scientists and policymakers were forced to face many questions that they would otherwise have ignored. So in spite of our human foibles, we are better for having dealt with these nontrivial matters, however ineffective and inefficient our efforts may have been. Now, with the scars of the war healed (and a few permanent disabilities) we may perhaps face the new challenges ahead with a modicum of wisdom. At least we should have learned to deal with one another better, and perhaps to understand and respect one another's views with less suspicion and worry. What lies ahead is more interesting and more important than what we have endured. We have much to do. With maturity, wisdom and self-restraint, and a bit of luck, no one will find it necessary to rush again to tell the King that the sky is falling.

The past decade and a half have brought such great advances in our understanding of the most fundamental properties of living cells that many describe it as a time of biological revolution. These advances were made possible by the development of several powerful new techniques that have allowed specific genes of many species to be isolated, characterized in detail, and inserted and expressed in other, more conveniently studied organisms. What followed as an immediate corollary of these methods was the development of an entirely new technology—biotechnology—the manipulation of organisms and the products they make for the solution of a variety of health and other problems.

Nanette Newell gives a brief account of these methods and the phenomena they have helped elucidate. She also provides a glimpse into the array of remarkable applications which these techniques have made possible.

APPENDIX

The Science and Technology of Gene Splicing

NANETTE NEWELL

As is brought out in other chapters in this volume, the invention of methods for manipulating the genetic material of all living things brought with it many questions. Could these techniques accidentally create genetic hazards? Will the course of evolution be changed in the laboratory, bypassing natural selection? Will these new methods be used to create the most nefarious biological weapons? Will they free us from the burdens of genetic diseases and cancer? Can biotechnology really solve the world's problems in food, energy, and pollution of the environment?

In order to appreciate the concerns over safety and the many philosophical questions that arise when one considers reordering the genetic material, it is necessary to understand the nature of the recombinant DNA technique itself. Elegant in its simplicity, it depends on knowledge of the structure and function of the genetic material and the properties of some unique enzymes. Recombinant DNA technology includes methods for the identification and isolation of some part of the genetic material (deoxyribonucleic acid, or DNA) from one organism, and its introduction, perhaps after some modification, into a new host, which can be a completely unrelated organism.[1] In this way, an entirely new functional genetic region can be put in place in a bacterium, plant, or animal, and it may include some chemically synthesized DNA as well as DNA from another, unrelated form of life. Use of the technology requires no large-scale investment in laboratory equipment nor specialized training of personnel, but the basic biological knowledge and the products that will result will certainly have a significant effect on the world in which we live.

This discussion includes a primer on the basic methodology of gene splicing and also offers an examination of the many uses to which these techniques, collectively known as biotechnology, are being applied. There are potential applications in virtually every industry that uses or produces chemicals, food, pharmaceuticals, energy, and to the treatment of wastes and chemical pollutants.

DNA and Genes

It is easier to understand how the recombinant DNA technique works when one is familiar with the structure of DNA (Figure 1). The fundamental chemical structure of DNA is identical in all organisms. The genetic information contained within the familiar double helix is completely defined by the linear order of four small subunits called nucleotide bases: adenine (A), guanine (G), cytosine (C), and thymine (T). These bases form the rungs of the DNA ladder. To form a rung, two bases, one from either side of the ladder, must come together in a specific fashion: A's can pair only with T's, and G's can pair only with C's. Therefore, the two sides of the ladder are not identical, but are complementary. The chemical nature of the complementary base pairing is very important to the function of DNA. The pairing across the ladder is specific, but not very strong, so that the DNA helix can "unzip" when necessary. Only in its unzipped form can the genetic information be copied and used.

Contained within the very long sequences of nucleotide bases are the units of heredity, knowns as genes. Usually, one gene contains the information needed to make one protein. Each gene is a certain ordered sequence of nucleotide bases with "signals" marking the beginning and end of the gene. The signals themselves are series of nucleotide bases. Thus, the sequences of the four bases, A, G, T, and C, depending on their order, contain the information for the synthesis of different proteins and the necessary signals for the correct use of the DNA. The genetic code—the sequences of nucleotide bases that specify how to build proteins—is the same for all organisms. On the other hand, the regulatory signal sequences can change from species to species. Thus, if DNA from one organism is inserted into the cells of another, that DNA will be propagated and expressed according to the rules in force in its new host. Animal DNA, once established in bacteria, will need bacterial signals in order to be functional in bacteria. This is one of the fundamental principles upon which all recombinant DNA experiments rest.

The DNA present in every cell of every living organism has the capacity

The Structure of DNA

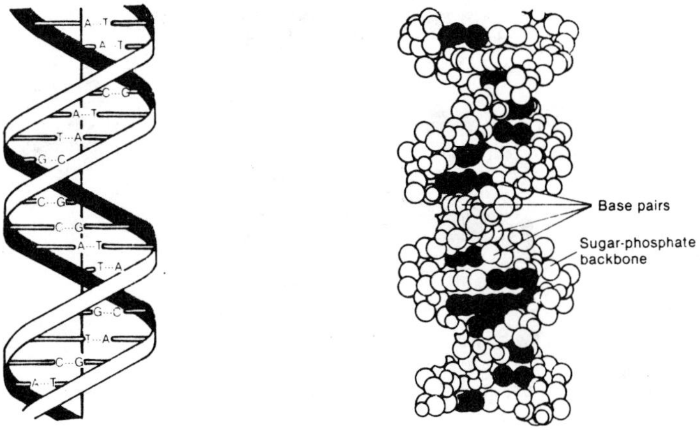

A schematic diagram of the DNA double helix.

A three-dimensional representation of the DNA double helix.

FIGURE 1. The structure of DNA. A schematic diagram of the DNA double helix and a three-dimensional representation of the DNA double helix. The DNA molecule is a double helix composed of two chains. The sugar-phosphate ribbons twist around the outside, with the paired bases on the inside serving to hold the chains together. Source: Office of Technonology Assessment.

to direct the functions of that cell. "Gene expression" refers to the way the genetic directions for particular cell functions are decoded and processed into the final functioning product, usually a protein (Figure 2). In the first step of this decoding, "transcription," part of the DNA double helix is unzipped, and the intermediate product, messenger RNA (mRNA) is synthesized. The mRNA, chemically very similar to DNA, is complementary to the section of unzipped DNA. In general, many copies of mRNA are made for each DNA region transcribed.

In the second step, the mRNA, after release from the DNA, becomes associated with the protein synthesizing machinery of the cell. There the sequence of nucleotide bases in the mRNA is decoded and translated into a particular protein. When that protein is no longer needed, both the protein and the mRNA for that protein are degraded. This mechanism allows both for the fine adjustment of the type and quantity of proteins in a cell and for keeping the DNA in a very stable and unaltered form.

The proteins perform the necessary functions of the cell. Some proteins are enzymes, which catalyze all biological reactions. Some proteins are structural and are found, for instance, in cell membranes and in virus particles. Structural proteins comprise the bulk of the protein, such as that in bone, cartilage, and muscle, found in nearly all members of the animal kingdom. Other proteins are regulators, such as those that bind to specific

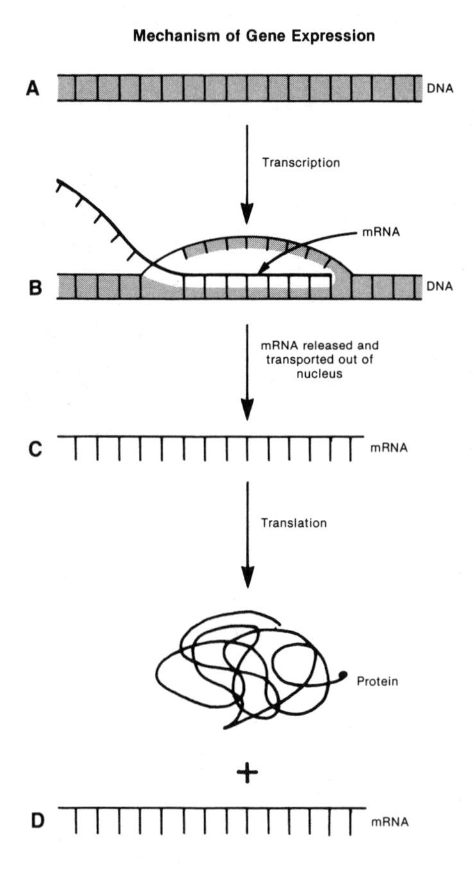

FIGURE 2. Mechanisms of gene expression. (A) A section of DNA containing a gene and the necessary regulatory signals. (The double-stranded structure is drawn without the helices for simplicity.) (B) During transcription the DNA is locally unzipped, and a complementary piece of RNA is synthesized. (C) The messenger RNA is released from the DNA and becomes associated with the cellular protein synthesizing machinery. (D) The mRNA is translated into protein. Source: Office of Technology Assessment.

locations on the DNA to control gene expression. Still other proteins have highly specialized functions. Hemoglobin, for example, is a protein that carries oxygen from the lungs to the rest of the tissues.

An ultimate goal of biologists is to understand in detail how humans function and how they interact with the ecosystem. Part of that understanding will come from the knowledge of how gene expression is regulated. All cells of any organism contain the full complement of DNA for that organism, but different genes are expressed in different cells. For example, it is the regulation of gene expression that causes liver cell to make the specialized products necessary for metabolizing food or for a brain cell to process information. The development of a human being from a single fertilized egg is precisely controlled by the regulation of genetic expression. There are hundreds, perhaps thousands, of genes that must be turned on and off at exactly the right time and in the right tissue for the accurate development of a higher organism to occur. Additionally, this control of gene expression must persist for the life of the organism. A breakdown in gene regulation is thought to be the cause of some diseases, most notably cancer. Therefore, knowledge about gene expression will not only satisfy man's inherent curiosity about himself, but may also bring with it the ability to understand and cure serious diseases.

The study of gene expression in higher organisms is extremely difficult because of the large number of genes present in any given cell. The amount of DNA present in each human (or higher animal) cell is approximately 3 billion nucleotide base pairs.[2] An average gene is about 1000 base pairs, or one-millionth of the total DNA. Since the study of one particular gene in the presence of all the others is very difficult indeed, one needs to isolate the gene, then make large quantities—clone it—so that the purified gene is available for study. This is exactly what rDNA technology can do.

General Strategy

In order to begin to purify, or clone, a gene, a "probe" for that gene is needed. A probe, in this case, is a molecule that can interact with the total cellular DNA in a specific manner to pick out the one gene of interest. A probe can be a piece of the gene, or the mRNA that is exactly complementary to the gene in question. The genes that can most easily be cloned are those from which one can get pure mRNA; unfortunately, these are limited to those genes that are expressed in very specialized cell types. For example, red blood cells make globin mRNA. Certain genes that are expressed abundantly only in specific tissues were the first cloned. Studies on these have proven enormously interesting, as discussed below, but as yet little is known about more common genes whose products every cell needs to carry on its routine housekeeping functions.

The protein products of genes offer an alternative procedure for obtaining a probe. Proteins cannot be used directly as probes, but because the genetic code is understood, one can work backward from a protein sequence and synthesize the corresponding DNA.[3] This synthetic DNA is then used as the probe for the gene sequences one wants to isolate. The process is difficult and tedious, but it is being automated and, in fact, will probably be the most widely used procedure for obtaining a probe. Synthetic sequences have been used for some time for the cloning of genes for small proteins such as hormones.[4]

In most gene cloning there are three parts to the experiment. First, the rDNA molecule must be introduced into the host cell to take advantage of the ability of host replication enzymes to make numerous copies of the DNA fragment. Finally, the cells that contain the DNA of interest (target DNA) must be selected or detected so that single clones of these cells can be isolated and grown in large quantities. The multiply-copied target DNA can then be isolated and purified for further study or for use in production processes.

Cloning of Complementary DNA

In this process, purified mRNA is copied to produce a stable DNA probe (Figure 3). (Messenger RNA is not used directly as a probe because it is not very stable and does not replicate itself.) When the mRNA, DNA nucleotide bases, and an enzyme (reverse transcriptase) are mixed together, a double-stranded helical structure is formed, one strand being the mRNA and the other strand being a complementary copy of DNA. The RNA is removed, and a double-stranded DNA structure is synthesized using another enzyme (DNA polymerase). This complementary DNA (cDNA) can now be used for the first step in gene cloning, the joining to a "vector," which will carry the target DNA.

The most important property of a vector is that it can replicate autonomously in an appropriate host, and, in the process, replicate the target DNA. The vector in this case is a bacterial plasmid, a small circular DNA molecule that occurs naturally in bacteria and that replicates independently from the bacterial chromosome.[5] The plasmid is linearized by cutting the DNA with a restriction enzyme, one of a class of enzymes that act as "molecular scissors." Restriction enzymes recognize a particular sequence of nucleotide bases and cut the DNA at that specific site.

Next, the cDNA must be joined to the plasmid. This is done by taking advantage of the property of complementary base pairing. The plasmid DNA is treated with another enzyme (terminal transferase) in the presence of a large amount of nucleotide base A to create single-stranded tails of nucleotides on either end of the DNA. The cDNA is treated in the same manner, only in the presence of T's. Now if the two "sticky-ended" DNAs are mixed together, the A's on the cDNA will base pair with the T's on the plasmids to form a circular piece of DNA that contains both the cDNA and the plasmid.

The plasmid containing the cDNA is then introduced into the host cell (bacteria in this case), which has been made receptive to the uptake of exogenous DNA. Very few of the bacterial cells take up the plasmid, so a good selection procedure is necessary to find them. Plasmids used as cloning vectors usually contain genes that confer antibiotic resistance on the cell into which they are introduced.[6] Therefore, if the bacterial cells are spread on

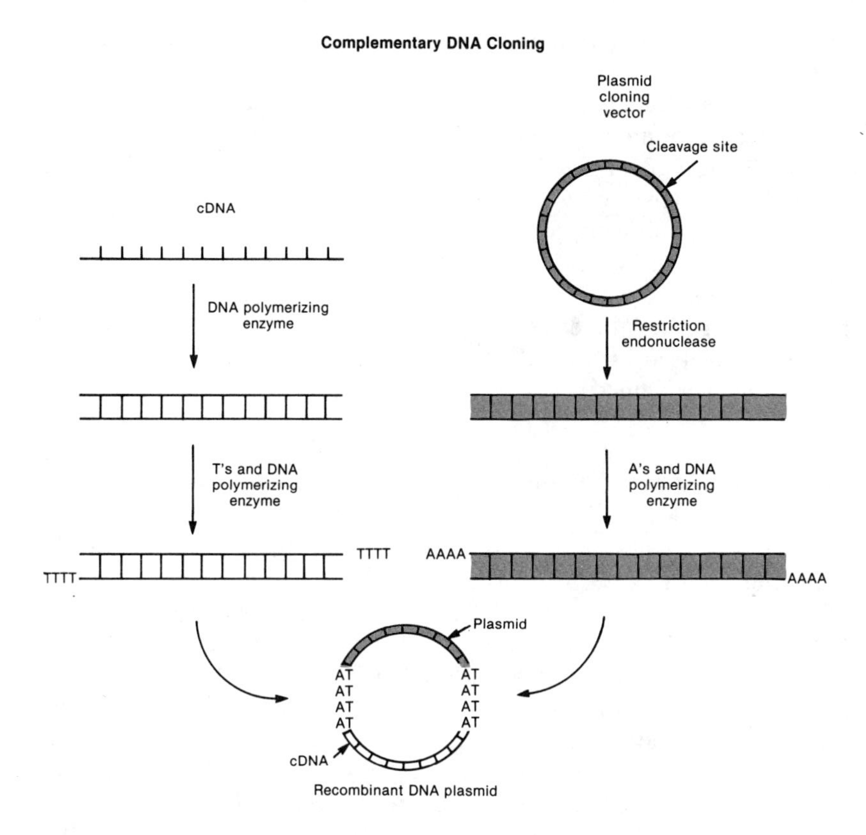

FIGURE 3. Complementary DNA (cDNA) cloning. The left column of the figure outlines the preparation for cloning of a cDNA from a mRNA. The right column outlines the preparation of the plasmid vector. (A plasmid is an extrachromosonal genetic element found in various strains of *E. coli* and other bacteria. The restriction endonuclease is an enzyme that acts as a "molecular scissors" and cuts the plasmid DNA circle at a specific site.) The molecule at the bottom of the figure schematically represents the completed cDNA plasmid ready for introduction into the host. (See text for discussion.) Source: Office of Technology Assessment.

agar (a solid nutrient medium) that contains the antibiotic, only those cells that contain the plasmid will grow.

It is also possible to identify the cDNA from a particular gene when the starting mRNA is a mixture from several genes. Rather than using a probe, the cloned cDNA is identified through the synthesis of at least part of the protein product of the cloned gene.[7] The cells containing the desired gene and producing its protein are detected by the binding of antibodies made specifically against this protein. Because antibodies bind only to small portions of a protein molecule, it is not necessary for the entire protein to be synthesized. This selection procedure allows the rapid identification of cloned genes in bacteria, which can then be grown in large quantity.

When the selected bacteria are grown on a larger scale in liquid culture, each cell can contain as many as a few hundred copies of the plasmid, and each milliliter of growth medium will contain about one billion cells. Therefore each milliliter of growth medium can contain over a trillion copies of the cDNA, a significant amplification. When the cDNA is purified and studied, much can be learned about the structure of the original mRNA. Other questions remain about the gene itself, however, such as the nature of the signal sequences at the beginning and end of the gene. The next step is to clone the gene itself by using the cDNA plasmid as a probe in a second kind of experiment explained below.

Total Cellular DNA Cloning

Cloning a single gene out of total cellular DNA (Figure 4) is more difficult because one must detect one particular DNA sequence out of a million. In order to have enough of a particular gene to detect, one usually amplifies the entire DNA by cloning.[8] Again the first step is the construction of recombinant DNA molecules. For rDNA experiments of this scale, one needs a more efficient vector system, usually a specially modified bacterial virus.[9] The viral DNA and the total cellular DNA are both cut with the same restriction enzyme to give DNA ends that all have the same terminal sequence. Not only do restriction enzymes cut DNA at specific sites, but the sites are pallindromic (the DNA base sequence reads the same forward on one strand as backwards on the other strand, e.g., GATC, CTAG), and the cuts are staggered. This enzyme action leaves overhanging, single-stranded ends that are self-complementary and complementary to the ends of other DNA molecules cut by the same enzymes (Figure 5). The mixture of vector and target DNA, whose ends have been allowed to form complementary base pairs, is treated with DNA ligase, an enzyme that seals the ends together. Then this DNA is put through a process that packages it into viral protein coats so that the DNA can be introduced into the bacteria.

The virus containing the rDNA is plated onto a bacterial "lawn" growing on agar. Every virus that produces an infection will kill the bacteria in its vicinity, giving an easily visible, clear spot in the normally opaque lawn. Each of these spots, or plaques, will contain about 10,000 copies of the viral rDNA, an adequate amplification. Each individual vector can accept about 1/150,000th of the total cellular DNA and, therefore, on the average, approximately 150,000 different plaques would need to be screened in order to find the gene of interest. One finds a specific sequence of DNA by using a specific cDNA probe that has had some radioactive atoms incorporated. The cDNA will blind only to its complementary sequences. Wherever the ra-

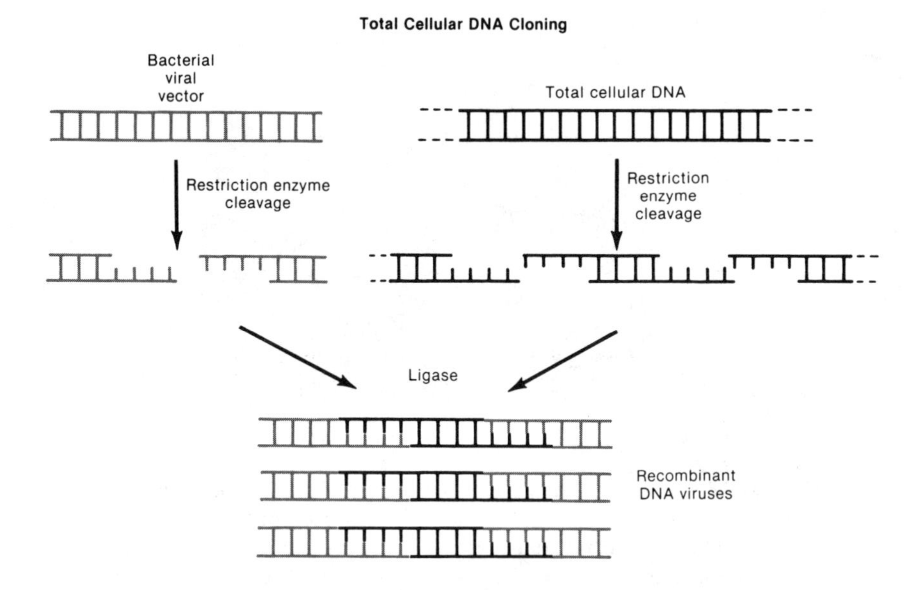

FIGURE 4. Total cellular DNA cloning. Both cellular and viral DNAs are cut with the same restriction enzyme to give complementary overhanging ends. The two cut DNAs are mixed and treated with ligase to give recombinant DNA molecules, part viral DNA and part cellular DNA. This DNA is subsequently packaged into viral protein coats prior to introduction into the host. Source: Office of Technology Assessment.

dioactive cDNA has bound to DNA in a plaque, it will expose a photographic film. The exposed spots on the developed film can be correlated with given plaques on the original plate, and the virus containing the DNA sequences of interest can be isolated and grown in large quantities. The DNA is easy to isolate from viruses for further study.

Recombinant DNA technology has at last allowed researchers to investigate the structure of genes of higher organisms in detail. One of the most intriguing discoveries to date is that most genes from higher organisms contain, in the midst of the information sequences, other sequences that do not contribute to the structure of the protein.[10] These sequences are transcribed along with the rest of the gene, but the "unnecessary" sequences are deleted before the RNA is translated into protein. At present the purpose of these intervening sequences is entirely unknown. Some researchers argue that because of nature's economy, the sequences will most likely be found to have a specific function. For instance, there is evidence that mRNA from some genes without intervening sequences is not found outside the nucleus of the cell,[11] so the processing of intervening sequences may be necessary for transport across the nuclear membrane. This idea correlates with the absence of intervening sequences in bacteria which have no nuclear membrane. There

Preparing "Sticky Ends":
The Use of Restriction Enzymes

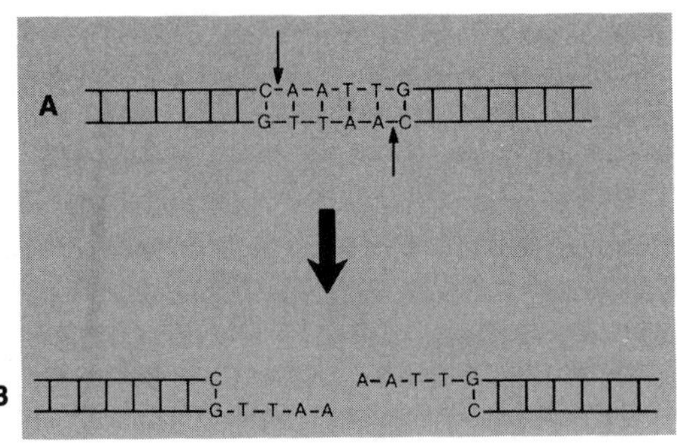

SOURCE: Office of Technology Assessment.

FIGURE 5. Restriction enzyme action. It was the discovery of these restriction enzymes that allowed the construction of rDNA molecules: specific fragments of DNA could be generated, and DNA from different species could be spliced together, because of the complementarity of the overhanging ends. Source: D. H. Hamer, K. D. Smith, S. H. Boyer, and P. Leder, "SV40 Recombinants Carrying Rabbit Beta-Globin Gene Coding Sequences," *Cell 17:* 725–735. D. Shortle, D. DiMaio, and D. Nathans, "Directed Mutagenesis," *Annual Review of Biochemistry,* 15 (1981): 265–294. (A) A six-base pair recognition site that is pallindromic, that is, it reads the same forward on the top strand as backward on the bottom strand. The arrows indicate the staggered cuts made by the enzyme. (B) The result of enzyme action is complementary, overhanging, single-stranded ends. Source: Office of Technology Assessment.

has also been speculation about the role of intervening sequences in evolution. Some researchers think these "extra" sequences may permit genetic variations to develop without jeopardizing the survival of the organism. A logical problem is solved for an evolutionary model that implies that greater molecular complexity will evolve over time. This model is difficult to reconcile with the notion that every genetic sequence codes for a necessary function.

Higher Organism Vector Systems

The cloning of a gene in a microorganism is sufficient if all one needs is a large quantity of that gene (with perhaps the addition of some control sequences) or its product. However, there are cases where it is necessary to clone genes from higher organisms into a vector system that can be grown in higher organism cells. This allows certain functions peculiar to the higher organism, such as the deletion of intervening sequences, to be studied.

Several vectors that function in higher organisms have been constructed; they have three important features in common.[12] The first feature, necessary for any cloning vector, is the presence of one or more restriction enzyme sites for the cloning of foreign DNA. Second, the vector must contain the genetic elements that permit it to grow independently and be detected in higher organism cells. The selection characteristic is usually a gene that encodes a protein necessary for the growth of the cells in a particular medium. That is, cells that do not contain the vector cannot grow when placed in a medium that contains the "selection" chemical; a situation exactly analogous to the antibiotic selection scheme described earlier for bacteria. Finally, these vectors must have genetic elements that allow them to be grown and detected in bacterial cells. This feature is important because growing large amounts of the cloned gene in a bacterial system is the fastest and cheapest amplification method currently available. Analogous to the situation in bacteria, DNA introduction into higher organism cells is inefficient and requires large quantities of DNA. Studies of gene expression in higher organisms using these vectors are now underway, and the next several years will surely see a large increase in our understanding of higher organism gene expression.

Industries Using Biotechnology

The current investigations into the structure of higher organism genes and the analysis of their expression is an exciting intellectual endeavor, but rDNA technology has also stimulated substantial industrial interest in biotechnology. Biotechnology, or the use of biological systems to perform useful chemical conversions, is not new, as is exemplified by the brewing, baking, and winemaking industries. In this discussion, however, biotechnology will refer to the use of the new genetic technologies to manipulate the genes of organisms for some useful industrial purpose.

Microorganisms can be modified to produce large amounts of the products of cloned genes or to perform some other biological task. For these purposes, it is necessary not only to isolate the gene(s) of interest, but also to achieve expression of that gene on a large scale.[13]

Two important points must be made about enhanced gene expression. First, recall that the regulatory signals that start and stop the decoding process are coded for in the DNA sequences on either side of the gene. Even though the genetic code is the same for all organisms, the regulatory signals vary among species and are quite different in bacteria and higher organisms. Therefore, achieving expression of a gene from one species in a cell of another species involves manipulation of the DNA surrounding the cloned gene so that the foreign gene product can be synthesized. Plasmids, known as expression vectors, have been constructed, using rDNA technology and restriction enzymes, that allow for genes to be inserted into specific places where the genes will be flanked by the appropriate regulatory sequences for the recipient host.[14] In some cases, over a million protein molecules can be produced per cell,[15] instead of the 10 to 100 copies of a protein normally found in a cell at any one time.

Second, it is important to remember that the structure of bacterial genes is different from genes of higher organisms, that is, only the higher organism genes contain nonessential intervening sequences. Bacteria have no mechanism for deleting them, so proper expression of the genes of higher organisms cannot occur in bacteria. Recall, however, that the cDNA is representative of the processed mRNA and contains only those sequences that contribute to the structure of the protein. Therefore cDNA clones are useful not only for probing the cellular genes, but also for encoding a protein product in microorganisms.

There are occasions when it may become necessary to synthesize gene products in higher organism cells. Many proteins in higher organism cells, especially those that are secreted, may be modified usually by the addition of sugar molecules. Bacteria cannot perform these modifications. For many proteins, the importance of the sugar groups are not known exactly, but if they are necessary for proper functioning, the genes will have to be cloned in a higher organism expression vector. Then they can be grown in yeast, animal, or plant cells to produce the functional protein product.

The commercial development of biotechnology is highly dependent on efficient gene expression because it is the proteins or the metabolites that are either the marketable products or that provide a cellular environment conducive to performing a task (e.g., bacteria that can degrade toxic wastes or a plant that can photosynthesize more efficiently). Also, with more sophisticated genetic manipulation techniques, it is now possible to change the sequence of nucleotide bases within a gene to make a protein product more suitable for a particular commercial application.[16]

Despite recent impressive gains in rDNA research, there are only a few rDNA products on the market, notably, human insulin[17] and a few animal vaccines. Nevertheless, the applications of rDNA technology to the needs and desires of people are potentially very large. The examples below give indications of the potential extent of the use of the technology.[18]

Human and Animal Health Care Products

Because the pharmaceutical industry already uses biologically produced products, it is particularly amenable to the use of rDNA technology.[19] Antibiotics, for instance, are complex compounds produced by microorganisms. Genetic manipulation could be used to increase yields or possibly to create new antibiotics.[20] Products currently produced biologically may be improved through the use of rDNA technology, but one of the first and most important uses of the technology is to produce in bacteria large quantities of compounds that are produced in only low amounts in higher organisms, therefore increasing the availability and lowering the cost. A notable example is interferon, a protein produced naturally in response to a viral infection and thought to have antiviral and anticancer properties. Because it has only been available in minute quantities of questionable purity, adequate clinical trials have not been possible. Now large quantities of pure, bacterially produced interferons are available,[21] and the unsubstantiated claims are being properly investigated.[22]

Another example is a large class of circulating regulatory proteins, collectively known as hormones. Insulin, used in the treatment of diabetes, is now isolated from pigs for use in humans. Some diabetics are allergic to porcine insulin, which differs slightly from human insulin. It may be desirable, therefore, for some diabetics to use human insulin. The gene for human insulin has been cloned and expressed in bacteria,[23] has undergone clinical trials, and is currently being marketed. Another hormone that has been even more difficult to obtain is human growth hormone, used in the treatment of dwarfism. Because its action is species specific, the only source to date has been from the pituitaries of deceased humans and shortages have been reported.[24] Bacterial production would solve the supply problem and presumably lower the cost. Bacterially produced growth hormone[25] is currently undergoing clinical trials.[26] Many more hormones and other small proteins will undoubtedly be produced using rDNA technology. For example, there is one class of small proteins, the enkephalins, that are produced in the nervous system and that appear to act as natural pain killers. Currently their genes are being cloned and studied.[27] Bacterially produced enkephalins could certainly be useful pharmacologically as potentially nonaddictive pain killers.

Inherited genetic diseases are usually due to a mutant gene that fails to produce a functional protein; this is the case in certain types of dwarfism. There are other instances where bacterially produced proteins could be used as treatment for genetic diseases. For example, people with hemophilia, a bleeding disorder, lack a certain protein in their blood. This protein could be made by a microorganism and used by hemophiliacs. Protein therapy is realistic for some blood disorders because the protein can be injected directly into the blood. However, there are many diseases characterized by protein deficiencies that require the protein to be present inside a particular cell type. Since proteins do not pass through cell membranes, protein therapy currently has limited use. There is another possible use of DNA technology, however, that will be addressed later.

The synthesis of viral vaccines is another promising application. Vaccines contain killed or attenuated viruses that stimulate an animal's immune system to produce antibodies against the viral surface proteins. When the animal is later exposed to the live virus, antibody defenses are already in place and can prevent an infection immediately. Traditional vaccines are produced in several steps. First, the virus in question is grown in tissue culture. Then, in one case, the virus is inactivated and purified for injection. Alternatively, nonvirulent viral mutants (attenuated strains) are selected in the laboratory, grown in tissue culture, and used for injection. There are many problems inherent in traditional vaccine production. Culture conditions may vary, resulting in nonuniform vaccines of varying potency. The inactivation process is critical in that the virus must be killed so that it does not cause an infection, but the surface proteins of the virus must remain intact enough to produce an immune response to the appropriate surface conformation. Sometimes the vaccine is not thoroughly inactivated and causes an outbreak of the disease it was designed to prevent. On the other hand, when the virus is killed, the integrity of the surface components may be destroyed, producing an ineffective vaccine.

Some vaccines do not exist at all because the virus of interest cannot be grown economically in tissue culture; rDNA technology may help circumvent such problems. Rabies and hepatitis are two diseases to which this new technology is being applied.[28] It is now possible to clone the genes that code for the viral surface proteins, and in some cases this has already been done. These cloned genes are placed in a bacterial expression vector, the viral surface proteins are produced in large amounts in bacteria, and when purified, they can be injected as a vaccine. More recently, it has been shown that small portions of viral surface proteins can also be effective as vaccines.[29]

Bacterial production solves many of the problems of traditionally produced vaccines. Because the growth of bacteria and the expression of genes within them is so well regulated, the resulting proteins are completely uniform and the vaccine potency is predictable and reliable. An unwanted infection is not possible because the genetic information of the infectious virus is not present in the vaccine. Growing the viral proteins, or portions of proteins, in bacteria, instead of growing the viruses in tissue culture, should ease scale-up of production and reduce the cost of vaccines below those produced by conventional methods.

New vaccines will be important for animals as well as humans. An early product of rDNA technology may very well be a vaccine for foot and mouth disease, a world-wide, devastating disease of cattle, swine, sheep, and goats. Traditional vaccines are available for this disease, but their effectiveness is limited. In fact, it is estimated that 44 percent of the cases of the disease in Europe in recent years have been caused by bad vaccines.[30] A vaccine produced by rDNA technology is undergoing field trials, and the preliminary results look promising.[31] The world market for this vaccine is estimated at 2.4 billion (monovalent) doses per year,[32] so presumably the animal health sector will see this as a desirable product to develop. In fact, the development of many animal vaccines is being viewed by industry with interest because of the large markets and the simpler and shorter regulatory requirements. There are now a commercial rDNA vaccine for scours, a severe diarrheal disease in newborn pigs and calves. A number of other viral diseases without effective vaccines that affect a variety of animals are also receiving attention.

Chemicals

Recombinant DNA technology can be applied to product and process development by the chemical industry.[33] This industry is largely dependent on petroleum as a substrate for the production of organic compounds. Now with the increased cost of oil and the ability to program bacteria to synthesize novel products, chemical companies are exploring the production of at least some organic compounds by biological techniques. It is possible to use biomass (cellulose, starch, organic solid wastes) as substrates for microorganisms, or for enzymes obtained from them, to produce many chemical compounds, including pesticides, amino acids, alcohols, and polymers. Theoretically, all organic compounds could be produced biologically, or at least most organic reactions could in principle be catalyzed by enzymes. But because many organic compounds in wide use today, such as Teflon, nylon, DDT, 2,4,5-T, or polyvinyl chloride, never existed in nature, no enzymes have arisen during the course of evolution to synthesize or degrade them. That is why many of these substances are "persistent" in the environment and cause serious pollution problems. Therefore, one of the compelling challenges to biotechnology is the design of totally new enzymes to catalyze some of these reactions. Molecular engineering can be expected to be one of the most important areas of biotechnology during the next decade, both for the synthesis of organic compounds and for the control of pollution.

Biological production methods are quite different from chemical ones and have different advantages and disadvantages. Biological or enzymatic processes generally use biomass as a chemical feedstock, which, in contrast to petroleum, is a renewable resource. This biomass, though, must always be pretreated using sometimes expensive thermal, chemical, or enzymatic processes. In contrast to the high temperatures and pressures used in many chemical production processes, the physical conditions for biological syntheses are mild. Organisms neither need nor will tolerate the high temperatures and pressures typical of chemical reactions using petroleum substrates. Biological conversions occur in dilute, aqueous solutions at temperatures and pressures compatible with life. This factor alone could reduce energy costs substantially.

For each step in a chemical synthesis, intermediate products must be separated out and purified. Microorganisms, on the other hand, can often be manipulated genetically so that they perform many of the intermediate steps in a complex synthesis within the cell and, therefore, purification is only necessary to obtain the final product. For example, strains of yeast are available that efficiently convert starch or cellulose to ethyl alcohol. Similarly, mixtures of enzymes, or a process involving sequential exposure to different immobilized enzymes, can result in a highly efficient conversion to the desired product.

Because of the lack of specificity and efficiency of chemical reactions, a synthesis may not completely convert the starting material and may produce many side products, some toxic, as well as the desired end product. During purification, these extraneous compounds must be disposed of or degraded. Converting them to nontoxic compounds is expensive and usually energy consuming; disposing of them in the environment can pose health hazards and cause ecological damage. Biological processes, which use enzymes as catalysts, yield fewer by-products and are extremely specific and efficient, sometimes producing essentially 100 percent conversion. Therefore many potential waste and disposal problems are eliminated. Biotechnology processes are generally considered nonpolluting and environmentally safe, unlike their chemical technology counterparts.

The most serious drawback to biological syntheses using microorganisms is the purification of the final product. The reactions take place in a dilute, aqueous environment, from which the product must be recovered, and the process of concentrating compounds is energy intensive. In some cases it may happen that the cost of purification exceeds the amount saved during the biological conversion steps of the process. Thus, not all biological or enzymatic processes offer cost advantages over conventional methods, and cost effectiveness of biological syntheses must be determined on a case-by-case basis, taking into account the points mentioned as well as the cost and availability of the starting materials. However, efforts to develop organisms and enzymes to work at high concentrations and better separations technology can be expected to lead to more economical processes.

Food Processing

The rDNA technology can be used in several different aspects of the food processing industry.[34] One application is the use of microorganisms to convert biomass to an edible product, either for human or animal consumption. This product, known as single cell protein (SCP), has been made in the past by conventional microorganisms and has been used as a protein supplement.[35] New technology allows organisms to be genetically manipulated to produce larger quantities of protein or to increase the rate at which the protein is produced. At present, the United States is putting very little effort into SCP production because it is not cost effective in this country; however, the Soviet Union considers the construction of SCP plants a high priority so that it can decrease its dependency on foreign sources of protein for animal food.[36] Essentially, SCP is a replacement for soybean protein, and soybeans are grown very efficiently in the United States. Organisms would have to be engineered to produce five to ten times as much protein as they do now before U.S. industries would find SCP worth producing commercially.[37]

Chemical conversions using microorganisms are not new to food processing, as exemplified by brewing, baking, and winemaking. During centuries of experience with yeast, workers have found extremely efficient strains to perform the needed reactions. Even so, rDNA technology is expected to have a significant impact on industries that use yeast.

Enzymes are also used in food processing to perform chemical conversions. Isolation of enzymes from their natural sources can be difficult and expensive, but if their genes can be isolated, they can be produced bacterially. An example is rennin, an enzyme necessary for the production of cheese. The highest quality rennin is produced naturally only in cow stomachs. The rennin gene has been cloned and expressed in a microorganism, which will make pure enzyme available in large quantities and reduce the dependence of the cheese industry on the beef industry.[38]

The food processing industry also makes chemicals used as food additives, and rDNA technology will undoubtedly be used in this aspect of food processing. The amino acid monosodium glutamate (MSG) is used extensively as a flavor enhancer, especially in Oriental cooking. The world market for MSG in 1980 was 600 million pounds.[39] Organisms that have been genetically manipulated to increase the production of MSG would have enormous commercial value. Another potentially important biological product is the synthetic noncaloric sweetener aspartame (a compound formed from two amino acids, phenylalanine and aspartic acid) that is coming into wide use. Many other potential applications of biotechnology in the food additive industry are for the production of chemicals to control spoilage and for flavors and fragrances, of polysaccharides that act as thickeners or gelling agents, and of enzymes that convert cellulose and starch to sugars.

Energy

Pharmaceuticals, chemicals, and food processing are obvious applications of the new technology, but the broader and potentially more important applications are in energy, agriculture, toxic waste degradation, and pollution control. Ethanol has been produced from biomass by the biological process of fermentation for several thousand years. Today, ethanol is a large-volume product of the chemical industry and is produced mainly by chemical means from petroleum. Recently, ethanol produced from biomass by fermentation has been used as a source of energy, mostly as a gasoline additive. This use of biomass is desirable not only because it reduces dependence on petroleum, but also because it displaces dirtier forms of energy, such as coal, and is a constructive use of biomass waste. Many different sources of biomass could be used to produce ethanol, as well as methane and hydrogen gas,[40] but the production of ethanol from cellulosic substances such as wood is of particular interest.

The energy input into the ethanol production process must be less than the energy obtained from the end product, otherwise it will not be a cost-effective energy source. For petroleum at current prices, present bioprocesses using biomass are for the most part not cost effective. However, genetically manipulated organisms should be able to increase the efficiency of the process.

When cellulose is enzymatically degraded to sugar, the sugar builds up, inhibits the functioning of the enzyme, and slows down the process. One solution to this problem has been to use fungus and yeast simultaneously. The one converts the cellulose to sugar and the other ferments the sugar to ethanol immediately.[41] Thus the sugar cannot inhibit the enzyme, and the process proceeds efficiently.

The most commonly used organism for ethanol fermentation, yeast, will tolerate a concentration of only about 12 percent ethanol. Recovery of ethanol from this dilute solution is expensive, and another desirable goal is to develop organisms with increased tolerance to ethanol.

Still another application of rDNA technology may be to develop organisms that can carry out fermentation at high temperatures.[42] If this is done, a combination fermentation and vacuum distillation process could allow the ethanol generated to be pumped off immediately.[43] Then the whole process could run continuously, never having an ethanol concentration high enough to kill the yeast. In addition, the high temperature would minimize contamination by other organisms.

Applications of rDNA technology can also be used to enhance the growth of plant matter to feed the energy production processes. Waste products such as wood chips can provide the major starting material for commercial ethanol production, but other rapidly growing plants producing large quantities of starch or cellulose may also be needed.

It is possible, and more economically feasible, to use grain as the starting material for ethanol production.[44] In fact, most of the ethanol produced in the United States by fermentation uses grain as a feedstock. But if grain ethanol contributed 10 percent of the automotive fuels used in the United States, over 40 percent of the annual U.S. grain harvest would be used for ethanol production.[45] On the other hand, the yield of ethanol per acre of trees can be as much as six times greater than that per acre of corn.[46] Still, the land use question of food versus fuel must be borne in mind.[47] Whether or not rDNA technology will be able to contribute to these aspects of energy production awaits much active investigation on several fronts.

Another use of microorganisms for providing energy is in the process of enhanced oil recovery. Often a well contains oil too viscous or at a pressure too low for pumping. It is hoped that microorganisms can be modified and placed in oil wells to produce compounds that will make the oil less viscous. Also, during the growth of these microbes, various gases such as carbon dioxide, methane, and nitrogen are produced, and these gases could repressurize the well. Thus, microorganisms might solve both problems of oil recovery at once. This difficult genetic feat has not yet been accomplished, but it is in the realm of possibility and is now the goal of several research groups.[48]

Waste Treatment and Pollution Control

Waste treatment and pollution control are also amenable to biological methods.[49] Human and animal waste, as well as most causes of pollution, are organic molecules. By the right chemical processes, they can be degraded to simple, harmless substances, primarily carbon dioxide and water. The "right" chemical process often involves burning, which is energy intensive and rarely 100 percent efficient. Bacteria, on the other hand, can be very efficient and thrive under mild conditions. Through classical genetics and the use of rDNA technology, it may be possible to develop bacteria that can completely digest waste compounds, including toxic ones. For instance, a new organism has been isolated that can live on 2,4,5-T, a pesticide and a component of Agent Orange.[50] Another was reported that degrades polychlorinated biphenyls (PCBs). It is possible that toxic waste dump sites will only be cleaned up efficiently and inexpensively by using microorganisms that have been genetically modified.

Plant Agriculture

Research on plants has been expanding at an unprecedented rate due to recent technical advances in the manipulation of the genetic material, the growth of plant cells in culture, plant cell fusion, and the capacity to regenerate whole plants from single cells.[51] The goals of current plant research include the enhancement of biological nitrogen fixation, photosynthesis, growth rate, yields per plant, nutritional quality, and of resistance to disease, pests, cold, and saline or alkaline soils. Recombinant DNA technology is being used to manipulate the soil bacteria, as well as to alter the plant genes themselves.

When discussing the manipulation of genes of higher organisms, it is important to keep two factors in mind. First, so little is known about the genetic organization and regulation in higher organisms that much more basic research needs to be done before accurate predictions about the use of rDNA technology in plants can be made. Consequently, some currently discussed potential applications are highly speculative. Second, because of the complexity of higher organisms, it is not known yet how to achieve the desired characteristic while maintaining high productivity. To date, there has always been a trade-off involved when changing the genetic compositions of higher organisms. For example, since the 1960s, genetic crosses have been described that improve the lysine content of corn. The adoption of these varieties has been slow, however, because the hybrids in which the high lysine genes are present give a slightly lower yield than standard varieties.[52]

Two technologies are crucial to the direct genetic manipulation of plants. One is the ability to culture plant cells and tissues under defined laboratory conditions. The other is the ability to regenerate a whole plant from a single cell. Every plant cell contains the genetic information necessary for its development into an adult plant. Genetic manipulations are significantly easier when done on isolated cells, and the regeneration capacity allows for the newly introduced genetic trait to be present in all cells of the plant when regenerated. It is important to remember, though, that currently, culture and regeneration conditions are known for only a few species. Laborious experimentation is required to find the laboratory conditions necessary for regeneration of each species, and the varieties that can be regenerated at present do not include the cereal grains, the group of plants used most commonly for food.

Even though little is known about gene organization and expression in plants, serious research efforts are underway to develop vector systems for introducing genes into plant cells.[53] Ultimately, the useful vectors will probably be of the type discussed earlier, those with the capacity to replicate in both bacterial and higher organism (in this case, plant) cells. Moreover, the ability of the vector to become attached to the plant's DNA would be a feature useful in assuring that every cell during regeneration receives a copy of the introduced gene.

One area making use of rudimentary vectors is nutritional improvement. Very few plants make "complete protein," which contains all the essential amino acids (those amino acids that humans require but cannot synthesize themselves). The isolation and amino acid analysis of some seed storage proteins from corn (zein)[54] and beans (phaseolin)[55] have been accomplished. Phaseolin is lacking methionine and zein lacks lysine, both essential. Some of the genes coding for the phaseolin proteins have been cloned[56] and inserted into plants, and expression has recently been achieved.[57] By the manipulation of the genetic code within a gene that codes for a major seed storage protein, the protein produced could be made complete and the plants more nutritious.

Plant research is also progressing rapidly in the identification and characterization of the genes responsible for disease and herbicide resistance. In most cases, resistances are coded for by one or two genes, making them accessible to genetic manipulation. A plant might be developed, for example, that was resistant to a particular herbicide; then the crop could be sprayed at any time to kill the weeds without damage to the crop.

Traits of most importance to farmers, such as high growth rate or high yield, are determined not by a single gene but by many, and these genes have yet to be identified. This fact points out the difficulty of using rDNA technology to address what are essentially physiologically defined characteristics. Much more research is needed to interweave the fields of plant physiology and plant molecular genetics. In the near future we shall only see applications of rDNA technology to genetically simple plant traits even though the field is moving quite rapidly.

Biotechnology can be used to genetically manipulate the microorganisms that associate with plants. Microorganisms are responsible for the fixing of nitrogen, an essential aspect of plant growth. Some microorganisms also produce pesticides and growth promotants. Since microorganisms are simpler than plants, the application of biotechnology to them may be seen in a shorter time.

In the midst of all the exciting and rapidly advancing molecular biology of plants, the problems and time frames associated with agricultural field trials must be kept in mind. Even after expression of a gene in a plant is accomplished, it still will take several years to do a field trial because, for most crops, there is only one growing season per year. Trials also are expensive and use a lot of land. Moreover, there have always been unexpected results in field trials, and testing genetically manipulated plants will probably be no exception.

Gene Therapy

The ability to manipulate genes in bacterial cells is very much a reality, although some of the applications will not be realized for some time. The use of rDNA technology for genes in plant cells is still in its infancy, but because of the ability to regenerate whole plants from single cells in culture, it is possible to imagine plants whose genes have been modified in the laboratory. An area of research that some consider the ultimate goal of genetic manipulation is gene therapy, i.e., the insertion of a foreign gene into an animal or human presumably to replace a dysfunctional gene. Several experiments have been done recently that suggest the feasibility of gene transplantation. For example, a cloned rabbit globin gene has been transferred into mice.[58] The experimenters injected purified rabbit globin gene into fertilized mouse eggs and implanted the eggs into foster mothers. The resulting mice were not only shown to have incorporated the foreign gene into their own DNA, but in a significant percentage of the cases the gene was actually expressed, and the investigators found rabbit globin in the mouse blood. The gene was stably incorporated since the mice could pass the rabbit gene to their offspring. In another more dramatic demonstration, fertilized mouse eggs were injected with the gene for rat growth hormone. The resulting mice, and their offspring, grew to up to twice the size of their mouse littermates.[59]

Experiments to date have been very preliminary, but more are being done, and these experiments will become increasingly sophisticated. As this technology advances and is combined with methods for mutating genes, its use in higher animals and man will become feasible. Replacing a dysfunctional gene is complicated by the fact that not only will the new gene have to contain the proper signals for its expression, but it must also overcome the deleterious effects of the dysfunctional gene which it is replacing. At this time, gene therapy can only be done with fertilized eggs or very young embryos in order to insure the gene's presence in many target cells at later stages of development. Possibly, we may learn how to incorporate genes into adult tissues, but this truly exciting aspect of recombinant DNA technology will most probably not be realized for a long time—if at all. Direct genetic intervention in animals has already been demonstrated and will be evident in the next several years, but intervention in humans in order to cure genetic disease will be complicated by ethical and legal issues as well as technical ones. In the meantime, the other applications of the powerful rDNA technology, in both university and industrial laboratories, will certainly result in increased knowledge about ourselves and our world and an improvement in the quality of our lives.

Notes

1. S. N. Cohen, "The Manipulation of Genes," *Scientific American* 233, no. 1 (1975): 24–33; P. Abelson, "A Revolution in Biology," *Science* 209 (1980): 1319–21; R. Wetzel, "Applications of Recombinant DNA Technology," *American Scientist* 68 (1980): 664–75; D. A. Hopwood, "The Genetic Programming of Industrial Microorganisms," *Scientific American* 245(3) (1981): 90–102; L. Stryer, *Biochemistry* (San Francisco: W. H. Freeman 1981); pp. 760-69; and A. Fisher, "Designer Genes," *Popular Science,* May 1982; pp. 92–94.
2. H. Ris and D. F. Kubai, "Chromosome Structure," *Annual Review of Genetics* 4 (1970): 263–94.
3. H. G. Khorana, "Total Synthesis of a Gene," *Science* 203 (1979): 614–25.
4. K. Itakura, T. Hirose, R. Crea, A. D. Riggs, H. L. Heyneker, F. Bolivar, and H. W. Boyer, "Expression in *Escherichia coli* of a Chemically Synthesized Gene for the Hormone Somatostatin," *Science* 198 (1977): 1056–63. See also R. Crea, A. Kraszewski, T. Hirose, and K. Itakura, "Chemical Synthsis of Genes for Human Insulin," *PNAS* 75 (1978): 5765–69; and D. V. Goeddel, D. G. Kleid,

F. Bolivar, H. L. Heyneker, D. G. Yansura, R. Crea, T. Hirose, A. Kraszewski, K. Itakura, and A. D. Riggs, "Expression in *Escherichia coli* of Chemically Synthesized Genes for Human Insulin," *PNAS* 76 (1979): 906–10.

5. R. C. Clowes, "The Molecule of Infectious Drug Resistance," in *Recombinant DNA,* offprints from *Scientific American* (San Francisco: W. H. Freeman, 1973), pp. 59–69. S. N. Cohen, F. Cabello, M. Casadaban, A. Cheng, and K. Timmis, "DNA Cloning and Plasmid Biology," in *Molecular Cloning of Recombinant DNA,* W. A. Scott and R. Werner, eds. (New York: Academic Press, 1977), pp. 35–56; R. L. Rodriguez, R. Tait, J. Shine, F. Bolivar, H. Heyneker, M. Betlach, and H. W. Boyer, "Characterization of Tetracycline and Ampicillin Resistant Plasmid Cloning Vehicles," in *Molecular Cloning of Recombinant DNA,* W. A. Scott and R. Werner, eds. (New York: Academic Press, 1977), pp. 73–84.
6. J. F. Morrow, "Recombinant DNA Techniques," in *Methods in Enzymology,* vol. 68, R. Wu, ed. (New York: Academic Press, 1979), 3–24.
7. D. Anderson, L. Shapiro, and A. M. Skalka, "*In situ* Immunoassays for Translation Products," in *Methods in Enzymology,* vol. 68, R. Wu, ed. (New York: Academic Press, 1979), pp. 428–36. See also L. Clarke, R. Hitzeman, and J. Carbon, "Selection of Specific Clones from Colony Banks by Screening with Radioactive Antibody," in *Methods in Enzymology,* vol. 68, R. Wu, ed. (New York: Academic Press, 1979), pp. 436–42; and H. A. Erlich, S. N. Cohen, and H. O. McDevitt, "Immunological Detection and Characterization of Products Translated from Cloned DNA Fragments," in *Methods in Enzymology,* vol. 68, R. Wu, ed. (New York: Academic Press, 1979), pp. 443–53.
8. F. R. Blattner, A. E. Blechl, K. Denniston-Thompson, H. E. Faber, J. E. Richards, J. L. Slightom, P. W. Tucker, and O. Smithies, "Cloning Human Fetal Gamma-Globin and Mouse Alpha-type Globin DNA: Preparation and Screening of Shotgun Collections," *Science* 202 (1978): 1279–84; T. Maniatis, R. C. Hardison, E. Lacy, J. Lower, C. O'Connel, D. Quon, G. K. Sim and A. Efstratiadis, "The Isolation of Structural Genes from Libraries of Eukaryotic DNA," *Cell* 15 (1978): 687–701.
9. K. Murray, "Applications of Bacteriophage Lambda in Recombinant DNA Research," *Molecular Cloning of Recombinant DNA,* W. A. Scott and R. Werner, eds. (New York: Academic Press, 1977), pp. 133–54. See also F. R. Blattner, B. G. Williams, A. E. Blechl, K. Denniston-Thompson, H. E. Faber, L. A. Furlong, D. J. Gunwald, D. O. Kiefer, D. D. Moore, J. W. Schumm, E. O. Shelton, and O. Smithies, "Charon Phages: Safer Derivatives of Bacteriophage Lambda for DNA Cloning," *Science* 196 (1977): 161–69.
10. R. Breathnach and P. Chambon, "Organization and Expression of Eucaryotic Split Genes Coding for Proteins," *Annual Review of Biochemistry* 50 (1981): 349–83.
11. D. H. Hamer, K. D. Smith, S. H. Boyer, and P. Leder, "SV40 Recombinants Carrying Rabbit Beta-Globin Gene Coding Sequences," *Cell* 17 (1980): 725–35.
12. P. Berg, "Dissection and Reconstruction of Genes and Chromosomes," *Science* 213 (1981): 296–304.
13. W. Gilbert and L. Villa-Komaroff, "Useful Protein from Recombinant Bacte-

ria," *Scientific American* 242, vol. 4 (1980): 74–94.

14. L. Guarente, T. M. Roberts, M. Ptashne, "A Technique for Expressing Eukaryotic Genes in Bacteria," *Science* 209 (1980): 1428–30.
15. D. G. Kleid, D. Yansura, B. Small, D. Dowbenko, D. M. Moore, M. J. Grubman, P. D. McKercher, D. O. Morgan, B. H. Robertson, and H. L. Bachrach, "Cloned Viral Protein Vaccine for Foot-and-Mouth Disease: Responses in Cattle and Swine," *Science* 214 (1981): 1125–28. See also D. Shortle, D. DiMaio, and D. Nathans, "Directed Mutagenesis," *Annual Review of Biochemistry* 15 (1981): 265–94.
16. D. Shortle, D. DiMaio, and D. Nathans, "Directed Mutagenesis."
17. I. S. Johnson, "Human Insulin from Recombinant DNA Technology," *Science* 219 (1983): 632–37.
18. U.S. Congress, *Impacts of Applied Genetics* Washington, D.C.: Office of Technology Assessment, 1981). See also A. L. Demain and N. A. Solomon, "Industrial Microbiology," *Scientific American* 245, no. 3 (1981): 66–75; and U.S. Congress, *Commercial Biotechnology: An International Analysis* (Washington, D.C.: Office of Technology Assessment, 1984).
19. Y. Aharonowitz and G. Cohen, "The Microbiological Production of Pharmaceuticals," *Scientific American* 245, no. 3 (1981): 140–52.
20. A. L. Demain, "New Applications of Microbial Products," *Science* 219 (1983): 709–14.
21. D. V. Goeddel, H. M. Shepard, E. Yelverton, D. Leung, R. Crea, A. Sloma, and S. Pestka, "Synthesis of Human Fibroblast Interferon by *E. coli*," *Nucleic Acids Res.* 8 (1980): 4057–74. See also D. V. Goeddel, E. Yelverton, A. Ullrich, H. L. Heyneker, G. Miozzari, W. Holmes, P. H. Seeburg, T. Dull, L. May, N. Stebbing, R. Crea, S. Maeda, R. McCandliss, A. Sloma, J. M. Tabor, M. Gross, P. C. Familletti, and S. Pestka, "A Human Leukocyte Interferon Produced by *E. coli* is Biologically Active," *Nature* 287 (1980): 411–16; and P. W. Gray, D. W. Leung, D. Pennica, E. Yelverton, R. Naharian, C. C. Simonsen, R. Derynck, P. J. Sherwood, D. M. Wallace, S. L. Berger, A. D. Levinson, D. V. Goeddel, "Expression of Human Immune Interferon cDNA in *E. coli* and Monkey Cells," *Nature* 295 (1982): 503–08.
22. T. Bodde, "Interferon: Will It Live Up to Its Promise?" *BioScience* 32 (1982): 13–15. See also D. Grady, "What Ever Happened to Interferon? *Discover* (March 1982): 83–85.
23. D. V. Goeddel et al., "Expressions in *Escherichia coli* of Chemically Synthesized Genes for Human Insulin," *PNAS* 76 (1979): 906–10.
24. P. Newmark, "Human Growth Hormone: Shortage Persists," *Nature* 294 (1981): 200–01.
25. D. V. Goeddel et al., "Direct Expression in *Escherichia coli* of a DNA Sequence Coding for Human Growth Hormone," *Nature* 281 (1979): 544–48. See also K. C. Olson, J. Fenno, N. Lin, R. N. Harkins, C. Snider, W. H. Kohr, M. J. Ross, D. Fodge, G. Prender, and N. Stebbing, "Purified Human Growth Hormone from *E. coli* Is Biologically Active," *Nature* 293 (1981): 408–11.
26. P. Newmark, "Human Growth Hormone." See also N. Angier, "Helping Chil-

dren Reach New Heights," *Discover* (March 1982): 35–40.

27. M. Comb, P. H. Seeburg, J. Adelman, L. Eiden, and E. Herbert, "Primary Structure of the Human Met- and Leu-enkephalin Precursor and Its mRNA," *Nature* 295 (1982): 663–66.

28. A. Anilionis, W. H. Wunner, and P. J. Curtis, "Structure of the Glycoprotein Gene in Rabies Virus," *Nature* 294 (1981): 275–78. See also J. C. Edman, R. A. Hallewell, P. Valenzuela, H. M. Goodman, and W. J. Rutter, "Synthesis of Hepatitis B Surface and Core Antigens in *E. coli*," *Nature* 291 (1981): 503–06; K. Hardy, S. Stahl, and H. Kupper, "Production in *B. subtilis* of Hepatitis B Core Antigen and of Major Antigen of Foot and Mouth Disease Virus," *Nature* 293, no. 3 (1981): 90–102; and P. Tiollais, P. Charnay, and G. N. Vyas, "Biology of Hepatitis B Virus," *Science* 213 (1981): 406–11.

29. R. A. Lerner, "Synthetic Vaccines," *Scientific American*, February 1983, pp. 66–74. See also E. Yelverton, S. Norton, J. F. Obijeski, and D. V. Goeddel, "Rabies Glycoprotein Analogs: Biosynthesis in *Escherichia coli*," *Science* 219 (1983): 614–20.

30. Howard Bachrach, U.S.D.A. Plum Island Research Facility, personal communication.

31. D. G. Kleid et al., "Cloned Viral Protein Vaccine for Foot-and-Mouth Disease."

32. Jerry Callis, U.S.D.A. Plum Island Research Facility, personal communication.

33. B. O. Palsson, S. Fathi-Afshar, D. F. Rudd, and E. N. Lightfoot, "Biomass as a Source of Chemical Feedstocks: An Economic Evaluation," *Science* 213 (1981): 513–17. See also D. E. Eveligh, "The Microbiological Production of Industrial Chemicals," *Scientific American* 245, no. 3 (1981): 154–78; and A. Hollaender, *Genetic Engineering of Microorganisms for Chemicals* (New York: Plenum Press, 1982).

34. A. H. Rose, "The Microbiological Production of Food and Drink," *Scientific American* 245, no. 3 (1981): 126–38.

35. K. J. Skinner, "Single-cell Protein Moves Toward Market," *Chemical and Engineering News* (May 1975): 24–26.

36. G. B. Carter, "Is Biotechnology Feeding the Russians?" *New Scientist* (23 April 1981): 216–18.

37. U.S. Congress, *Impacts of Applied Genetics*.

38. K. Nishimori et al., "Cloning in *Escherichia coli* of the Structural Gene of Prorennin, the Precursor of Calf Milk-Clotting Enzyme Rennin," *Journal of Biochemistry* 90 (1981): 901–04.

39. U.S. Congress, *Impacts of Applied Genetics*.

40. H. G. Schlegel and J. Barnea, *Microbial Energy Conversion* (New York: Pergamon Press, 1977). See also H. P. Gregor and T. W. Jeffries, "Ethanolic Fuels from Renewable Resources in the Solar Age," *Biochemical Engineering*, ed. by W. R. Vieth et al. (New York: New York Academy of Sciences, 1979); U.S. Congress, *Energy from Biological Processes*, vols. 1 and 2 (Washington, D.C.: Office of Technology Assessment, 1980); H. Bungay, *Energy, the Biomass Option*

(New York: Wiley, 1981); W. Tyner, "Biomass Energy Potential in the United States of America," *Mazingira* 5, no. 1 (1981): 44–53; W. Worthy, "Cellulose-to-Ethanol Projects Losing Momentum," *Chemical and Engineering News,* 7 December 1981, pp. 35–42; and H. DeYoung, "Ethanol from Biomass: The Quest for Efficiency," *High Technology* 2, no. 1 (1982): 61–69.

41. P. J. Blotkamp, M. Takagi, M. S. Pemberton, and G. H. Emert, "Enzymatic Hydrolysis of Cellulose and Simultaneous Fermentation to Alcohol," *AIChE Symposium Series No. 181*, vol. 74 (1978): 85–90. See also G. H. Emert and R. Katzen, "Gulf's Cellulose to Ethanol Process," and D. B. Rivers and G. H. Emert, "Integration of Enzyme-Catalyzed Cellulose Degradation with Ethanol Production," *Proceedings World Bio-Energy '80* (1980).
42. J. A. Miller, "Hot Bug for Energy," *Science News,* 3 November 1979, p. 217. See also J. Wiegel and L. G. Ljungdahl, "Isolation and Characterization of a New Extreme Thermophilic Anaerobic Bacteria," *Abstracts of the Annual Meeting of the American Society of Microbiology* (1979): 105; and J. G. Zeikus, "Thermophilic Bacteria: Ecology, Physicology, and Technology," *Enzyme and Microbial Technology* 1 (1979): 243–52.
43. J. L. Glick, "Fundamentals of Genetic Technology and Their Impact on Energy Systems," 9th Energy Technology Conference, Washington, D.C., 1982, personal communication.
44. H. Bungay, *Energy, the Biomass Option.*
45. P. Rothberg and M. Segal, "Energy from Solid Wastes and Bioconversion," Issue Brief 74064, Congressional Research Service, Library of Congress (1981).
46. J. D. Ferchak and E. K. Pye, "Utilization of Biomass in the U.S. for the Production of Fuel as a Gasoline Replacement," *Farm and Fuel Produced Alcohol: The Key to Liquid Fuel Independence* (Joint Economic Committee, U.S. Congress): 29–53.
47. G. G. Marten, "Land Use Issues in Biomass Energy Planning," *Resources Policy,* (March 1982), pp. 65-74.
48. R. Myers, "Bugs Boost Texas Oil Output," *The Energy Daily* 10, no. 38 (26 February 1982). See also W. R. Finnerty and M. E. Singer, "Microbial Enhancement of Oil Recovery," *Bio/Technology*, March 1983, pp. 47–54.
49. H. Kobayashi and B. E. Rittman, "Microbial Removal of Hazardous Organic Compounds," *Environmental Science and Technology* 16 (1982): 170A–83A.
50. S. T. Kellogg, D. K. Chatterjee, and A. M. Chakrabarty, "Plasmid-Assisted Molecular Breeding: New Technique for Enhanced Biodegradation of Persistent Toxic Chemicals," *Science* 214 (1981): 1133–35.
51. "The Second Green Revolution," *Business Week*, 25 August 1980, pp. 92D–92L. See also W. Brill, "Agricultural Microbiology," *Scientific American* 245 no. 3 (1981): 198–215; E. C. Cocking, M. R. Davey, D. Pental, and J. B. Power, "Aspects of Plant Genetic Manipulation," *Nature* 293 (1981): 265–70; "Breeding Better Crops," *The Economist,* 5 December 1981, pp. 106–07; E. Leepson, "Advances in Agricultural Research," *Editorial Research Reports* 1, no. 19 (1981): 371–88; B. Tamarkin, "The Growth Industry," *Forbes,* 2 March 1981, pp. 90–94; A. Fisher, "High-Tech Plants," *Popular Science*, May 1982, pp. 92–

94; E. Lerner, "World Lags in Applying Crop Technology," *High Technology* 2, no. 1 (1982): 80–86; and J. Trevis and A. Bertelsen, "Genetic Engineering: Promise for Agricultural Industries," *Feedstuffs*, 1 February 1982, pp. 32–41.

52. E. T. Mertz, "Case Histories of Existing Models," *Genetic Improvements of Seed Proteins* (Washington, D.C.: National Academy of Sciences, 1976).

53. M. D. Chilton, D. A. Tepfer, A. Petit, C. David, F. Casse-Delbart, and J. Tempe, "*Agrobacterium rhizogenes* Inserts T-DNA into the Genomes of the Host Plant Root Cells," *Nature* 295 (1982): 432–34. See also F. A. Krens, L. Molendijk, G. J. Wullems, and R. A. Schilperoort, "*In vitro* Transformation of Plant Protoplasts with Ti-plasmid DNA," *Nature* 296 (1982): 72–74.

54. J. L. Fox, "More Nutritious Corn Aim of Genetic Engineering," *Chemical and Engineering News,* 7 December 1981, pp. 31–33.

55. S. M. Sun, J. L. Slightom, and T. C. Hall, "Intervening Sequences in a Plant Gene—Comparison of the Partial Sequence of cDNA and Genomic DNA of French Bean Phaseolin," *Nature* 289 (1981): 37–41.

56. Ibid.

57. T. C. Hall, "Modification of Crop-Seed Proteins," in *Abstracts of Papers of 148th National Meeting,* A. Herschman, ed. (Washington, D.C.: American Association for the Advancement of Science, 1982), p. 44.

58. T. E. Wagner, P. C. Hoppe, J. D. Jollick, D. R. Scholl, R. L. Hodinka, and J. B. Gault, "Microinjection of a Rabbit Beta-Globin Gene into Zygotes and Its Subsequent Expression in Adult Mice and Their Offspring," *PNAS* 78 (1981): 6376–80.

59. R. D. Palmiter et al. "Dramatic Growth of Mice That Develop from Eggs Microinjected with Metallothionein-Growth Hormone Fusion Genes," *Nature* 300 (1982): 611–15.

Glossary

Aerobic: Growing only in the presence of oxygen.

Alcohol: Any of a series of organic compounds that have one or more -C-O-H groups. Ethanol (grain alcohol) is a simple alcohol composed of two carbon atoms, six hydrogen atoms, and an oxygen atom: H_3C-H_2C-O-H. Methanol (wood alcohol), with only one carbon atom, is the simplest alcohol: H_3C-O-H.

Amino acids: The building blocks of proteins. There are twenty common amino acids; they are joined together in a strictly ordered "string" that determines the character of each protein.

Anaerobic: Growing only in the absence of oxygen.

Antibody: A protein component of the immune system in mammals; found in the blood.

Antigen: A large molecule, usually a protein or carbohydrate, which when introduced into the body stimulates the production of an antibody that will react specifically with the antigen.

Bacteria: A major class of prokaryotes.

Bacteriophage: A virus that multiples in (parasitizes) bacteria. Bacteriophage lambda is commonly used as a vector in recombinant DNA experiments.

Biomass: Plant and animal material.

Biome: A community of living organisms in a major ecological region.

Biosynthesis: The production of a chemical compound by a living organism.

Biotechnology: The collection of industrial processes that involve the use of biological systems. For some industries, these processes involve the use of genetically engineered microorganisms.

Carbohydrates: The family of organic molecules consisting of simple sugars such as glucose and sucrose, and sugar chains (polysaccharides) such as starch and cellulose.

Catalyst: A substance that enables a chemical reaction to take place under milder than normal conditions (e.g., lower temperatures). Biological catalysts are enzymes; nonbiological catalysts include metallic complexes.

Cell fusion: The process whereby two or more cells come together to become a single cell.

Cell lysis: Disruption of the cell membrane allowing the breakdown of the cell and exposure of its contents to the environment.

Cellulose: A polysaccharide (sugar chain) composed entirely of glucose units linked end to end; it constitutes the major part of cell walls in plants.

Chimera: An individual composed of a mixture of genetically different cells.

Chromosomes: The thread-like nuclear components of a cell that contain the hereditary material of the organism. They are composed of DNA and protein and contain most of the cell's DNA.

Clone: A group of genetically identical cells or organisms asexually descended from a common ancestor. All cells in the clone have the same genetic material and are exact copies of the original.

Cloning vector: Small DNA molecules—usually plasmid or virus—into which foreign DNA has been inserted. After reintroduction into the microorganisms, the DNA molecules are replicated (cloned), producing billions of identical copies of the hybid DNA.

Conjugation: The one-way transfer of DNA between bacteria in cellular contact.

Containment: Provision of a sealed laboratory hood or other enclosure, or of a separate, sealed laboratory complex, that permits access to experimenters while preventing escape of particles, bacteria, or other harmful materials. P-1 to P-4 physical containment levels provide for increasingly stringent controls to confine organisms containing recombinant DNA molecules so that the potential hazard to laboratory workers, persons outside the laboratory, and the environment is minimal. These containment levels are described in *Federal Register,* vol. 48, no. 106 (June 1983), Notices, pp. 24569–24574.

Cytogenetics: A branch of biology that deals with the study of heredity and variation by the methods of both cytology (the study of cells) and genetics (the study of heredity and variation in plants and animals).

Cytokines: Any of a group of plant hormones that elicit certain plant growth and development responses, especially by promoting cell division.

Cytoplasm: The protoplasm (mainly water, proteins, fatty substances, carbohydrates, and inorganic salts) of a cell, external to the cell's nuclear membrane.

DNA (deoxyribonucleic acid): The genetic material found in all living organisms. Every inherited characteristic has its origin somewhere in the code of an individual's complement of DNA.

DNA polymerase: An enzyme that synthesizes DNA fibers.

DNA vector: A vehicle for transferring DNA from one cell to another.

Dominant gene: A characteristic whose expression prevails over alternative characteristics for a given trait.

Endotoxins: Complex molecules (lipopolysaccharides) that are an integral part of the cell wall and are released only when the integrity of the cell is disturbed.

Enkephalins: Naturally occuring molecules composed of amino acids that are produced in the nervous system and appear to act as painkillers.

Enterotoxins: Poisonous substances produced by microorganisms that cause some types of food poisoning.

Enzyme: A functional protein that catalyzes a chemical reaction. Enzymes control the rate of metabolic processes in an organism; they are the active agents in the fermentation process.

Escherichia coli (E. coli): A bacterium that commonly inhabits the human intestine. It is a favorite organism for many microbiological experiments.

Eukaryote: A complex, compartmentalized cell characterized by its extensive internal structure and the presence of a nucleus containing the DNA. All multicellular organisms are eukaryotic. The simpler cells, the prokaryotes, have much less compartmentalization and internal structure.

Exotoxins: Proteins produced by bacteria that are able to diffuse out of the cells; generally more potent and specific in their action than endotoxins.

Fermentation: The biochemical process of converting a raw material such as glucose into a product such as ethanol.

Fibroblast: A cell that gives rise to connective tissues.

Fibroma: A tumor composed primarily of fibrous connective tissue.

Fungi: Plants (usually single-cell) that must depend entirely on other organisms for food. Many fungi are destructive pathogens; others are important in the production of foods and pharmaceuticals.

Gene: The hereditary unit; a segment of DNA that codes for a specific protein.

Gene amplification: Production of extra copies of a gene within a cell.

(display)

Gene expression: The manifestation of the genetic material of an organism as specific traits.

Gene mapping: Determining the relative locations of different genes on a particular chromosome.

Genetic code: The biochemical basis of heredity consisting of codons (base triplets along the DNA sequence) that determine the specific amino acid sequence in proteins and that are the same for all forms of life studied so far.

Genetic engineering: A technology used at the laboratory level to alter the hereditary apparatus of a living cell so that the cell can produce more or different chemicals, or perform completely new functions. These altered cells are then used in industrial production.

Genome: The basic chromosome set of an organism—the sum total of its genes.

Genotype: The genetic constitution of an individual or group.

Germplasm: The total genetic variability available to an organism, represented by the pool of germ cells or seed.

Hormones: The "messenger" molecules of the body that help coordinate the actions of various tissues; they produce a specific effect on the activity of cells remote from the point of origin of the hormone.

Hydrocarbon: Any organic compound that is composed only of carbon and hydrogen.

Immunoproteins: All proteins that are part of the immune system (including antibodies, interferon, and cytokines).

Immunotoxin: Any antitoxin in blood that confers immunity against a disease.

Interferon: A protein, produced by intact animal cells when they are infected with viruses, that acts to inhibit viral reproduction and to induce resistance in host cells.

In vitro: Outside the living organism and in an artificial environment.

In vivo: Within the living organism.

Leukocytes: The white cells of blood.

Lipids: Water-insoluble biomolecules such as cellular fats and oils.

Messenger RNA: Ribonucleic acid molecules that serve as a guide for protein synthesis.

Metabolism: The sum of the physical and chemical processes involved in the maintenance of life and by which energy is made available.

Mitochondria: Structures in higher cells that serve as the "powerhouse" for the cell, producing chemical energy.

Monoclonal antibodies: Antibodies derived from a single source or clone of cells that recognize only one kind of antigen.

Mutant: An organism whose visible properties with respect to some trait differ from the norm of the population due to mutations in its DNA.

Mutation: Any change that alters the sequence of bases along the DNA, changing the genetic material.

Mycoplasmas: Organisms that are gram-negative, generally nonmotile, nonsporing bacteria lacking a true cell wall. They are distinguished by their sterol requirement for growth.

Mycotoxin: A poison produced by a fungus.

Myxoma: A tumor composed of mucinous connective tissue.

Myxomatosis: A virus disease of rabbits that produces skin lesions and swelling of mucous membranes.

Nif genes: The genes for nitrogen fixation present in certain bacteria.

Nucleic acid: A polymer composed of DNA or RNA subunits.

Nucleotides: The fundamental units of nucleic acids. They consist of one of the four bases—adenine, guanine, cytosine, and thymine (uracil in the case of RNA)—and its attached sugar-phosphate group.

Organic compounds: Chemical compounds based on carbon chains or rings, which contain hydrogen, and which also may contain oxygen, nitrogen, and various other elements.

P-1–P-4 containment: *See* containment.

Pathogen: A specific causative agent of disease.

Peptide: Short chain of amino acids.

Phage: *See* bacteriophage.

Phenotype: The visible properties of an organism that are produced by the interaction of the genotype and the environment.

Phosgene: A highly toxic, colorless gas (carbonyl chloride) that has been used in warfare.

Plasmid: Hereditary material that is not part of a chromosome. Plasmids are circular and self-replicating. Because they are generally small and relatively simple, they are used in recombinant DNA experiments as acceptors of foreign DNA.

Polymer: A giant molecule made from the union of simple molecules (monomers). A long-chain carbohydrate (e.g., cellulose or starch) is a polymer containing molecules of simple sugars linked together.

Prokaryotes: Bacteria (and certain algae) whose DNA is not organized into chromosomes or surrounded by a nuclear membrane. The DNA is a simple strand in the cell.

Protein: A linear polymer of amino acids. Proteins are the products of gene expression and are the functional and structural components of cells.

Protoplast: A cell without a wall.

Protoplast fusion: A means of achieving genetic transformation by joining two protoplasts or joining a protoplast with any of the components of another cell.

Recessive gene: Any gene whose expression is dependent on the absence of a dominant gene.

Recombinant DNA: The hybrid DNA produced by joining pieces of DNA from different sources in vitro.

Restriction enzyme: An enzyme within a bacterium that recognizes and degrades DNA from foreign organisms, thereby preserving the genetic integrity of the bacterium. In recombinant DNA experiments, restriction enzymes are used to recognize particular sequences of molecular units in DNA and act as tiny biological scissors to cut up "foreign" DNA before it is recombined with the DNA of the cloning vector.

Reverse transcriptase: An enzyme that can synthesize a single strand of DNA from a messenger RNA, the reverse of the normal direction of processing genetic information within the cell.

Rickettsia: A group of insect parasites that are prokaryotic and somewhat intermediate between bacteria and viruses. Rickettsia causes a number of serious diseases in humans, including Rocky Mountain spotted fever and typhus.

RNA (ribonucleic acid): In its three forms—messenger RNA, transfer RNA, and ribosomal RNA—it assists in translating the genetic message of DNA into the finished protein.

Senescence: The complex of aging processes that lead to death.

Somatic cell: One of the cells composing parts of the body (e.g., tissues, organs) other than a germ cell.

Tissue culture: An in vitro method of propagating healthy cells from tissues, such as fibroblasts from skin.

Transduction: The process by which foreign DNA becomes incorporated into the genetic complement of the host cell.

Transformation: The transfer of genetic information by DNA separated from the cell.

Vector: A transmission agent. A DNA vector is a self-replicating DNA molecule that transfers a piece of DNA from one host to another. *See* cloning vector.

Virus: An infectious agent that requires a host cell in order for it to replicate. It is composed of either RNA or DNA wrapped in a protein coat.

Yeast: A variety of fungi.

Index